PROTEINS

Function

Albert Light

Department of Chemistry
Purdue University

PRENTICE-HALL, INC., Englewood Cliffs, New Jersey

Library of Congress Cataloging in Publication Data

LIGHT, ALBERT
Proteins: structure and function.

Includes bibliographies.
1. Proteins. I. Title. [DNLM: 1. Proteins—Analysis. QU55 L723p 1974]
QD431.L58 547'.75 73-13613
ISBN 0-13-731711-5

PRENTICE-HALL INTERNATIONAL, INC., *London*
PRENTICE-HALL OF AUSTRALIA, PTY. LTD., *Sydney*
PRENTICE-HALL OF CANADA, LTD., *Toronto*
PRENTICE-HALL OF INDIA PRIVATE LIMITED, *New Delhi*
PRENTICE-HALL OF JAPAN, INC., *Tokyo*

10 9 8 7 6 5 4 3 2 1

Printed in the United States of America

CONTENTS

PREFACE

This book provides a treatment of protein chemistry that can accompany or closely follow a course in general biochemistry. The material is written at a level which presupposes an adequate undergraduate preparation in chemistry.

Today the techniques of protein chemistry appear in numerous biochemical and biological studies that are sometimes remote from the specific study of proteins, and that may include such physiological processes as protein biosynthesis, enzymology, transport, genetics and evolution, vision, and muscular contraction. More and more often, the specialist in a specific biological study reaches a stage in his research where he must examine his system at the molecular level. When this occurs, it is clear often that protein chemistry is playing an essential and critical part in these studies.

The major topics of protein chemistry are reviewed in the text to provide a useful starting point for those who wish to gain an insight into the current methods of studying the structure of proteins and the relationship of structure to function. The elucidation of the amino acid sequence of the protein molecule is the particular aspect of protein chemistry that exists as the underlying theme. I believe that a knowledge of the sequence is necessary for us to appreciate and understand the chemical, physical, and biological properties of the molecule.

The major part of the text deals with the experimental approach that is followed by most investigators who wish to perform a sequence analysis. It is clear that the sequence of the protein alone is insufficient to explain the structure and properties of the molecule, and consequently, the three-dimensional structure is also introduced. In the last chapter, we treat the subject of structure and function and deal with specific amino acid residues that have a unique role in the biological activity of the protein. Sequence variation among similar proteins and also in a single protein as a function of species is discussed as an excellent example of how sequence studies provide information on structure-function relationships, on the three-dimensional structure, and on evolution.

At the end of each chapter, a limited number of references are presented. Generally these are textbooks and major review articles although a few specific journal references are also included.

The author made no attempt to provide a complete survey of the subject. Instead, a judicious choice of topics, timely examples, and pertinent methods were selected and offered to the reader. Although the coverage is limited in scope, it should be clearly recognized as sufficient to give the reader an appreciation of those procedures that are usually employed for studies of protein structure. In many places, the inclusion of certain topics and the exclusion of other subjects reflect the personal prejudices of the author as well as his experience with these topics. Apologies are given to those investigators whose work would be covered by other writers but are omitted here for the sake of simplifying the exposition.

It is a pleasure to acknowledge the help and participation of many colleagues and friends in the preparation of this book. I am particularly grateful to Drs. S. R. Suskind, W. B. Jakoby, and J. L. Markley for their critical comments, suggestions, and corrections of errors of the entire manuscript. I also wish to thank Miss Barbara Schatzley for typing part of the manuscript during working hours and my wife, Toby, for her typing at all other hours. My wife and both children, Pam and Audrey, gave me the needed incentive and the encouragement that helped to complete the writing of the book.

ALBERT LIGHT

PROTEINS

Structure and Function

Basic Principles of Protein Chemistry

ONE

Our understanding of life processes will be placed on a more rational basis as we acquire a deeper appreciation of the fundamental chemical and physical properties of various cell constituents. Early attempts to unravel the mechanism of specific physiological processes concentrated on the role of proteins, and these efforts are continuing at an ever-accelerating pace.

Proteins are the principal constituent of cells; they are ubiquitous throughout nature and essential to life itself. The dominant role played by proteins in living systems can be appreciated with the selection of a few examples that illustrate the variety of physiological properties associated with these molecules. All chemical reactions in living systems are catalyzed by enzymes, protein molecules that are catalysts either by themselves or when complexed with small molecules. Proteins participate in such diverse and necessary processes as respiration, muscular contraction, active transport of cellular constituents, electrical transmission, and in the expression and perpetuation of genetic characteristics. Proteins function as hormonal regulators of metabolism, as storage depots for certain molecules, in immunological defense mechanisms, and as structural elements.

Protein chemistry is the subject that deals with the chemical and physical properties of proteins and the relationship of their structure to function. The initial studies on the chemical properties of proteins can be traced to the early 1800s when Mulder first used the term "protein" to describe those substances of living systems, such as gelatin, fibrin, and casein, which he recognized as complex nitrogeneous substances. A characteristic feature of protein molecules is their nitrogen content (about 14 to 17 percent by weight). However, it required a recognition that amino acids served as the basic structural unit of proteins before rapid progress could be made on a systematic approach to the study of protein structure.

The discovery of glycine by Braconnot in 1820 was the first of the amino acids to be identified as a constituent of proteins. Glycine was obtained from gelatin after hydrolysis with sulfuric acid, neutralization of the acid with lime, and crystallization of the amino acid from the concentrated syrup. Although many other amino acids were identified soon thereafter, new discoveries were still to be made in the twentieth century. It is still very likely that other amino acids will be discovered as minor constituents of particular types of protein structures.

AMINO ACIDS

Structural units of proteins

Our present tally of the amino acids that are found as constituents of proteins is given in Fig. 1-1, where the structures are presented as amino acid residues. Since the amino acids differ in their side chain substituents, the "amino acid wheel" focuses attention on the chemical nature of these groups. Consequently, the structures are grouped as aliphatic, aliphatic alcohol, sulfur, imino acid, acidic, amide, basic, and aromatic. The three-letter abbreviation is commonly used when protein sequences are reported and the one letter abbreviations are useful when related protein sequences are compared.

Stereoisomerism

Although the two-dimensional structure of an amino acid correctly describes the chemical constituents, one must also consider its three-dimensional arrangement in order to appreciate fully the size and positions of the atoms attached to the alpha carbon. Figure 1-2 shows a ball and stick arrangement and a space-filling model of **D**- and **L**-alanine. The tetrahedral structure of the carbon atom clearly positions each of the groups in space and fixes the relative order of one group to the next. Amino acids found as constituents of proteins are of the **L** configuration. We return later in this chapter to the spatial relations of amino acids when we consider the arrangement of amino acid residues in polypeptide chains.

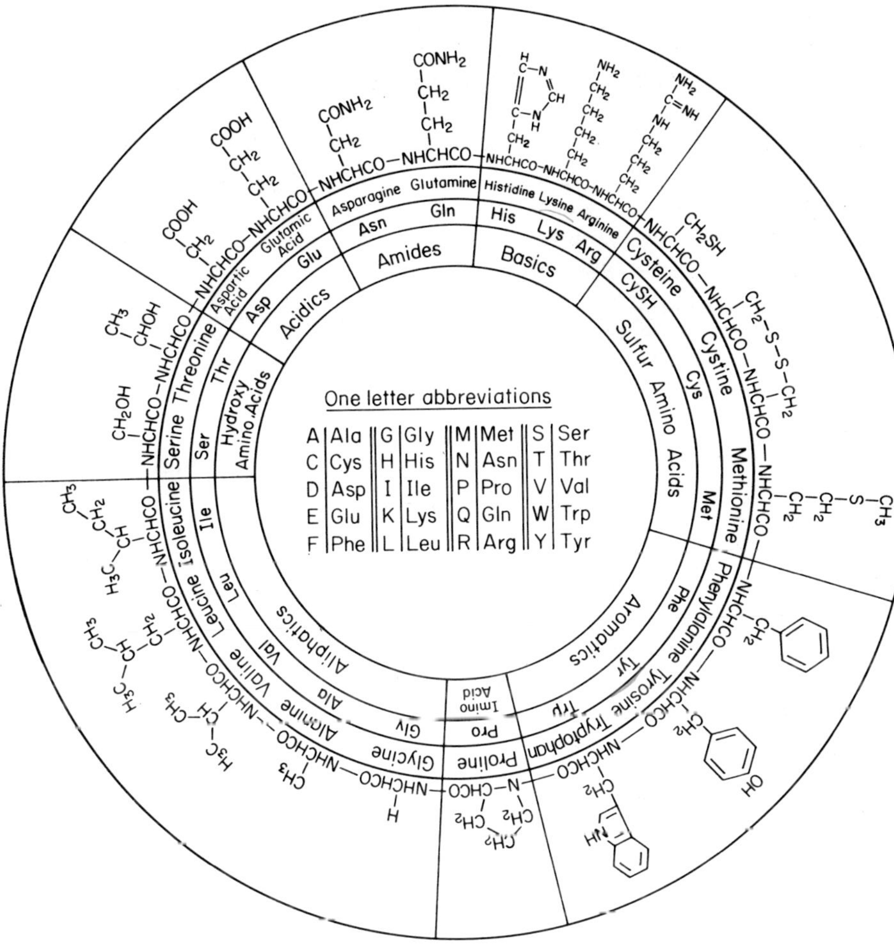

Fig. 1-1. The "amino acid wheel." The structures of the amino acid residues are shown as they normally occur in a polypeptide chain. The names of the amino acid residues and their three-letter and one-letter abbreviations are given below the structures. The amino acids are placed in groups (aliphatics, acidics, basics, etc.) based on the properties of the side chain substituents.

THE PEPTIDE BOND

Amino acids are covalently linked to one another in a protein by peptide bonds, as first suggested by Fischer and Hoftmeister in 1902. The peptide bond hypothesis is now supported by a variety of experimental observations and is accepted as the unified basis for the structure of all proteins. The polypeptide chain of a protein can be represented, as shown below, for a structure of $(n + 2)$ amino acid residues.

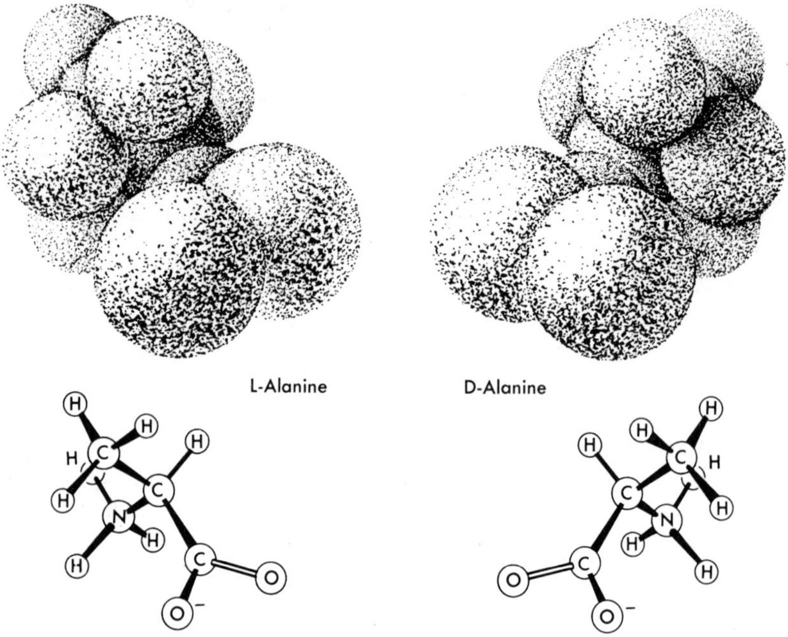

Fig. 1-2. The stereoisomers of the amino acid alanine. The space-filling model (upper half) is a good representation of the relative size of the atoms and the steric relationships of the groups of the α-carbon atom. The ball-and-stick model (lower half) closely resembles the two-dimensional structure and it serves as a guide to the identification of the atoms of the space-filling model. ("From *General Chemistry*, 3rd Edition, by Linus Pauling. W. H. Freeman and Company, copyright © 1970.")

The α-carboxyl of one amino acid and the α-amino of a second amino acid (or the imino group of proline) participate in the formation of an amide bond, linking the two amino acids. The repeating structure of amino acids linked to one another by peptide bonds is referred to as the backbone of the polypeptide chain. The type of side chain will vary from one residue to another in the polypeptide, so that the nature of the side chains gives the protein molecule its distinctive properties and functional behavior.

$$NH_2\overset{R}{\overset{|}{C}}HCO\text{–}(NH\overset{R}{\overset{|}{C}}HCO)n\text{–}NH\overset{R}{\overset{|}{C}}HCOOH$$

PRIMARY STRUCTURE OF PROTEINS

The amino acid sequence of a protein, which is also referred to as the primary structure, is a complete description of the molecule in organic chemical terms, equivalent to the structural information available for small organic molecules. In later chapters the strategy and methods of sequence analysis of a protein will be discussed. For the present it is sufficient to note that the application of sequence studies has been spectacularly successful and that the complete sequences of many different proteins are now known.

It is clear from the sequence analysis of a large number of different proteins that each protein has a unique structure, that the sequence is characteristic of the species, and that, in a few cases, the structure may also be specific for a particular organ or tissue. Furthermore, the sequence of a protein is faithfully reproduced as cells divide. Thus we can refer to bovine insulin, or, more correctly, to proinsulin, which is the precursor of the biologically active hormone, as a specific structure that will be elaborated by the islet cells of the pancreas of all normal cows (Fig. 1-3). Furthermore, the sequences of bovine insulin and insulin from other mammalian species were almost identical, usually differing in a single amino acid residue, with the change located in the A chain at residues 8, 9, or 10. These similarities soon prompted others to expand these studies, and a present view of the amino acid sequences of insulins from a number of species is shown in Table 1-1. Although amino acid variations occur among different species, 18 positions apparently never change. The study of the sequence of a protein from different species represents one of the approaches now used to relate structure to function (Chap. 11).

The primary structure of a protein provides us with a description of the organic-chemical nature of the molecule, but it fails to deal with those features of the protein that describe the conformation of the polypeptide chain. Other experimental approaches must be used before one can arrive at a true picture of the protein molecule.

SECONDARY STRUCTURE OF PROTEINS

Physical measurements of various types strongly suggest that globular proteins should not be represented as fully extended polypeptide chains but rather as compact, highly folded structures. A complete description of the protein molecular must therefore include a detailed analysis of the conformation of the polypeptide chain and the position and bonding of side chain substituents. The secondary structure of a protein is concerned with a description of the three-dimensional arrangement of the polypeptide chain—the conformation of the polypeptide backbone.

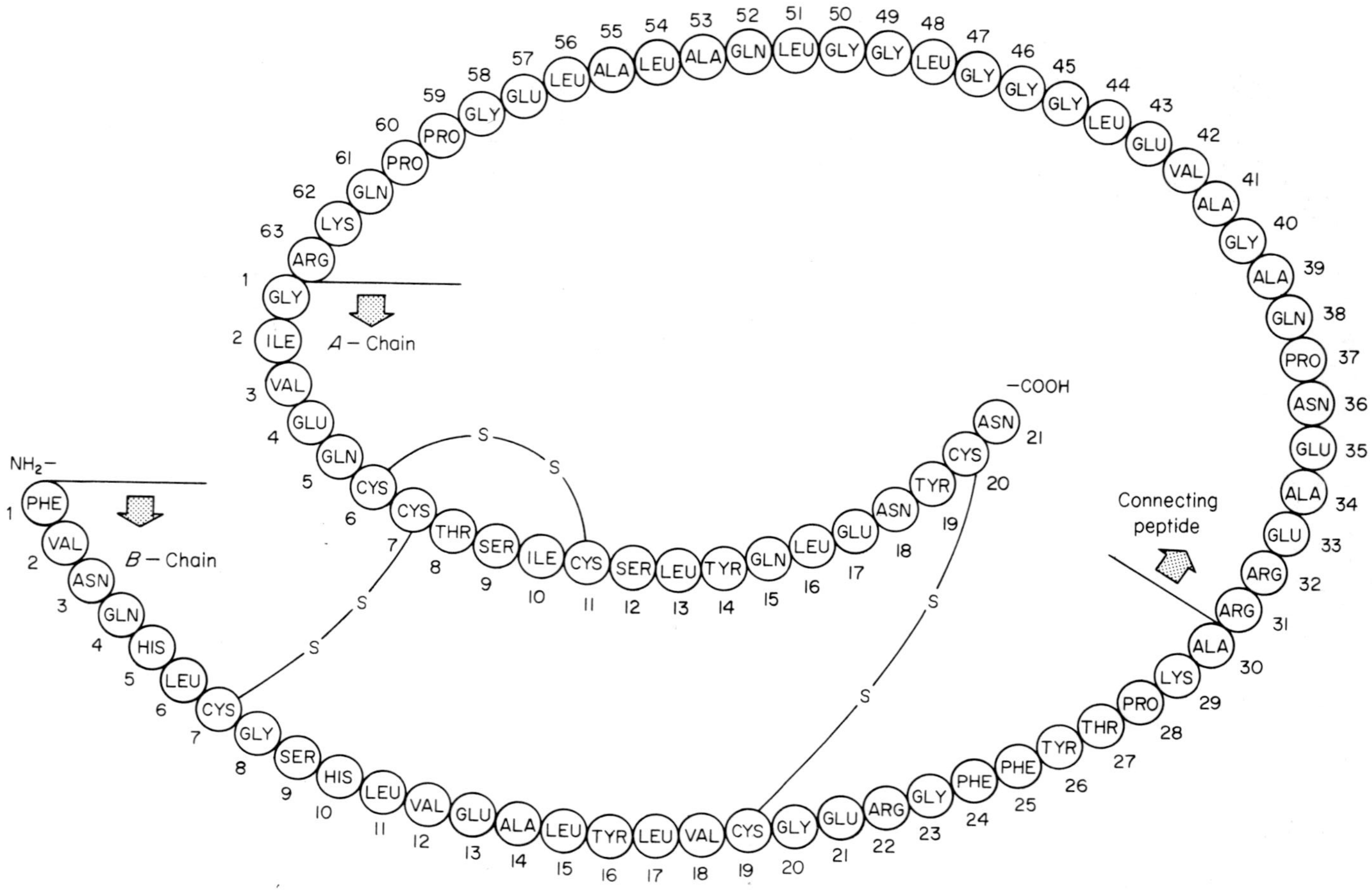
NH$_2$–
B – Chain
A – Chain
–COOH
Connecting peptide
S
S

Fig. 1-4. The bond lengths and bond angles of a fully extended polypeptide chain. [From L. Pauling, R. B. Corey, and H. R. Branson, *Proc. Natl. Acad. Sci. U. S.*, 37, 205 (1951).]

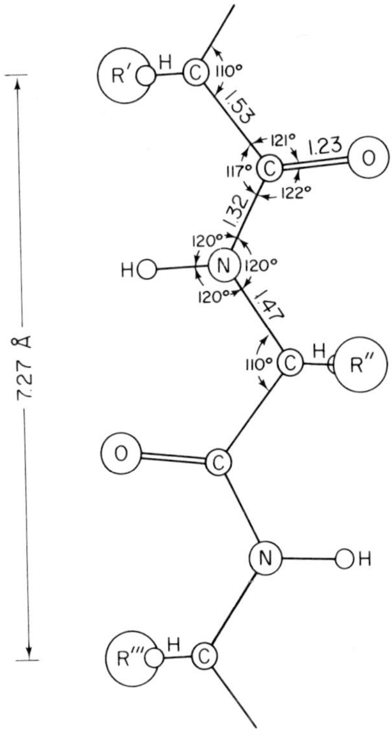

Planar nature of the peptide bond

The structure of the peptide unit was established by Pauling and Corey from x-ray diffraction studies of individual amino acids and of several di- and tripeptides (Fig. 1-4). The bond angles and bond lengths of the α-carbon atom are typical of a tetrahedral carbon. However, the C–N bond distance of the peptide bond is intermediate between a single and double bond. The partial double-bond character of the peptide bond helped Pauling and Corey to recognize the planar nature of the peptide bond. The four substituents of the peptide

---

Fig. 1-3. Proinsulin is a single polypeptide chain elaborated by the islet cells of the pancreas and serves as the precursor molecule of the hormone insulin. Limited proteolysis of proinsulin causes the loss of the "connecting peptide," leading to the two-chain structure–the A chain and B chain–typical of insulin molecules. The sequences of insulin of different species are given in Table 1-1. [From R. E. Chance, R. M. Ellis, and W. W. Bromer, *Science*, **161**, 165 (1968). Copyright © 1968 by the American Association for the Advancement of Science and R. E. Chance, *Diabetes*, **21**, Supplement 2, 461 (1972).]

TABLE 1-1

The amino acid differences of insulin of different species[a,b]

															B Chain							
								(38) ↑												(51) ↑		
										1										2		
	1	2	3	4	5	6	7	8	9	0	1	2	3	4	5	6	7	8	9	0	1	2
Human	–	F	V	N	Q	H	L	C	G	S	H	L	V	E	A	L	Y	L	V	C	G	E
Rat and mouse 1	–	F	V	K	Q	H	L	C	G	P	H	L	V	E	A	L	Y	L	V	C	G	E
Rat and mouse 2	–	F	V	K	Q	H	L	C	G	S	H	L	V	E	A	L	Y	L	V	C	G	E
Guinea pig	–	F	V	S	R	H	L	C	G	S	N	L	V	E	T	L	Y	S	V	C	Q	D
Rabbit	–	F	V	N	Q	H	L	C	G	S	H	L	V	E	A	L	Y	L	V	C	G	E
Elephant	–	F	V	N	Q	H	L	C	G	S	H	L	V	E	A	L	Y	L	V	C	G	E
Horse	–	F	V	N	Q	H	L	C	G	S	H	L	V	E	A	L	Y	L	V	C	G	E
Pig[c]	–	F	V	N	Q	H	L	C	G	S	H	L	V	E	A	L	Y	L	V	C	G	E
Bovine	–	F	V	N	Q	H	L	C	G	S	H	L	V	E	A	L	Y	L	V	C	G	E
Sheep, goat	–	F	V	N	Q	H	L	C	G	S	H	L	V	E	A	L	Y	L	V	C	G	E
Sei whale	–	F	V	N	Q	H	L	C	G	S	H	L	V	E	A	L	Y	L	V	C	G	E
Chicken, turkey	–	A	A	N	Q	H	L	C	G	S	H	L	V	E	A	L	Y	L	V	C	G	E
Cod	M	A	P	P	Q	H	L	C	G	S	H	L	V	D	A	L	Y	L	V	C	G	D
Toadfish 1	M	(A.	P.	P)	Q	H	L	C	G	S	H	L	V	D	A	L	Y	L	V	C	G	D
Toadfish 2	M	(A.	P.	P)	Q	H	L	C	G	S	H	L	V	D	A	L	Y	L	V	C	G	D
Angler fish	V	A	P	A	Q	(H.	L)	C	(G.	S.	H.	L.	V.	D.	A.	L.	Y.	L.	V)	C	(G.	D)
Tuna 2	V	A	P	P	Q	H	L	C	G	S	H	L	V	D	A	L	Y	L	V	C	G	D
Bonito	–	A	A	N	(P.	H.	L)	C	(G.	S.	H)	L	V	E	A	L.	Y	L.	V	C	(G.	E)
COMMON						H	L	C	G			L	V			L	Y		V	C		
ALTERNATIVES	–	F	V	N	Q					S	H			E	A			L			G	E
	M	A	P	P	R					P	N			D	T			S			Q	D
	V		A	K	P																	
				S																		
				A																		

[a]Adapted from M. O. Dayhoff (Ed.), *Atlas of Protein Sequence and Structure*, Vol. **5**, National Biomedical Research Foundation, Washington, D. C., 1972.

[b]The three disulfide bonds of insulin are indicated by the arrows above the residue numbers.

[c]Dog, Sperm Whale, Finback Whale.

bond are almost always of the trans configuration, such that the oxygen of the carbonyl group and the hydrogen of the nitrogen lie diagonally across from one another.

The planar nature of the peptide bond imposes constraints on the number of possible orientations that may occur when two or more amino acids are joined in a polypeptide chain Since the α-carbon atom acts as a swivel point between two peptide planes, the orientations are described by the angle of rotation allowed at the C_α–N bond, the angle ϕ, and by the angle of rotation at the C_α–C bond,

TABLE 1-1 (cont.)

The amino acid differences of insulin of different species[a,b]

																			A Chain										
														(42)↑	(8)↑				(37)↑									(20)↑	
							3										4										5		
3	4	5	6	7	8	9	0	1	2	3	4	5	6	7	8	9	0	1	2	3	4	5	6	7	8	9	0	1	2
R	G	F	F	Y	T	P	K	T	G	I	V	E	Q	C	C	T	S	I	C	S	L	Y	Q	L	E	N	Y	C	N
R	G	F	F	Y	T	P	K	S	G	I	V	D	Q	C	C	T	S	I	C	S	L	Y	Q	L	E	N	Y	C	N
R	G	F	F	Y	T	P	M	S	G	I	V	D	Q	C	C	T	S	I	C	S	L	Y	Q	L	E	N	Y	C	N
D	G	F	F	Y	I	P	K	D	G	I	V	D	Q	C	C	T	G	T	C	T	R	H	Q	L	E	S	Y	C	N
R	G	F	F	Y	T	P	K	S	G	I	V	E	Q	C	C	T	S	I	C	S	L	Y	Q	L	E	N	Y	C	N
R	G	F	F	Y	T	P	K	T	G	I	V	E	Q	C	C	T	G	V	C	S	L	Y	Q	L	E	N	Y	C	N
R	G	F	F	Y	T	P	K	A	G	I	V	E	Q	C	C	T	G	I	C	S	L	Y	Q	L	E	N	Y	C	N
R	G	F	F	Y	T	P	K	A	G	I	V	E	Q	C	C	T	S	I	C	S	L	Y	Q	L	E	N	Y	C	N
R	G	F	F	Y	T	P	K	A	G	I	V	E	Q	C	C	A	S	V	C	S	L	Y	Q	L	E	N	Y	C	N
R	G	F	F	Y	T	P	K	A	G	I	V	E	Q	C	C	A	G	V	C	S	L	Y	Q	L	E	N	Y	C	N
R	G	F	F	Y	T	P	K	A	G	I	V	E	Q	C	C	A	S	T	C	S	L	Y	Q	L	E	N	Y	C	N
R	G	F	F	Y	S	P	K	A	G	I	V	E	Q	C	C	H	N	T	C	S	L	Y	Q	L	E	N	Y	C	N
R	G	F	F	Y	N	P	K/		G	I	V	D	Q	C	C	H	R	P	C	D	I	F	D	L	Q	N	Y	C	N
R	G	F	F	Y	N	P	K/		G	I	V	E	Q	C	C	(H.	R.	P)	C	(D.	I)	F	D	L	Q	S	Y	C	N
R	G	F	F	Y	N	S/			G	I	V	E	Q	C	C	(H.	R.	P)	C	(D.	K)	F	D	L	Q	S	Y	C	N
R	(G.	F.	F.	Y.	N)	P	K/		(G.	I.	V.	E.	Q)	C.	C	(H.	R.	P)	C	(N.	I.	F.	D.	L.	Q.	N.	Y)	C	N
R	G	F	F	Y	N	P	K/		G	I	V	E	Q	C	C	H	K	P	C	N	I	F	D	L	Q	N	Y	C	N
R	G	F	F	Y	Q	P	K/		G	I	(H.	Z.	Z.	C.	C	H.	K.	P.	C.	B.	I)	F	Z	L	Z	D	Y	C	N
	G	F	F	Y					G	I			Q	C	C				C					L			Y	C	N
R					T	P	K	A			V	E				T	S	I		S	L	Y	Q		E	N			
D					N	S	M	S			H	D				H	G	P		D	I	F	D		Q	S			
					S			T								A	R	V		N	R	H							
					I			D									K	T		T	K								
					Q												N												

the angle ψ (Fig. 1-5). The angles of rotation—namely, the orientation of adjacent planar structures—are dictated, in general, by the attempt to minimize steric hindrance and to maximize stability between neighboring groups. The carbonyl group of the peptide bond is bulky, and it must be placed as far away as possible from neighboring carbonyl groups and α-carbon substituents. These considerations limit the number of degrees of freedom available to neighboring residues, thus placing a limitation on the allowed rotations of the angles ϕ and ψ.

An important stabilizing element in the polypeptide structure is the potential for hydrogen bonding. The polar nature of the peptide bond enhances the

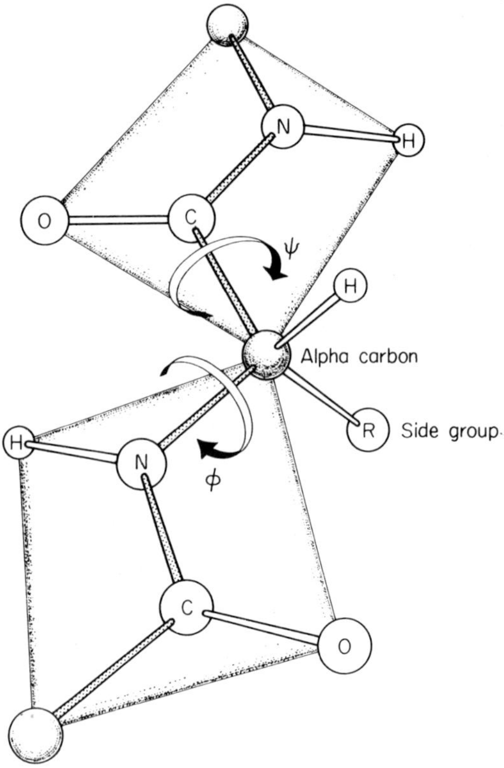

Fig. 1-5. Perspective drawing of a section of polypeptide chain representing two polypeptide units. The angles of rotation for a pair of planar peptide units are shown as ϕ, for the rotation of the C_α–N bond, and ψ, for the rotation of the C_α–C bond. The tetrahedral structure of the C_α serves as a swivel point. The limits of a residue are indicated by dashed lines. The chain is shown in a fully extended conformation ($\phi = \psi = 180°$) and the residue is of the L-configuration. The designation of the dihedral angles of ϕ and ψ for the conformation of a polypeptide follows the recommendations of the IUPAC-IUB Commission on Biochemical Nomenclature. [From the IUPAC-IUB, CBN Document. Reprinted from *Biochem.*, **9**, 3471 (1970). Copyright © 1970 by the American Chemical Society. Reprinted by permission of the copyright owner.]

possibility of interactions between the carbonyl group of one peptide bond and the amide group of another.

$$\rangle C{=}O \cdot\cdot\cdot H{-}N\langle$$

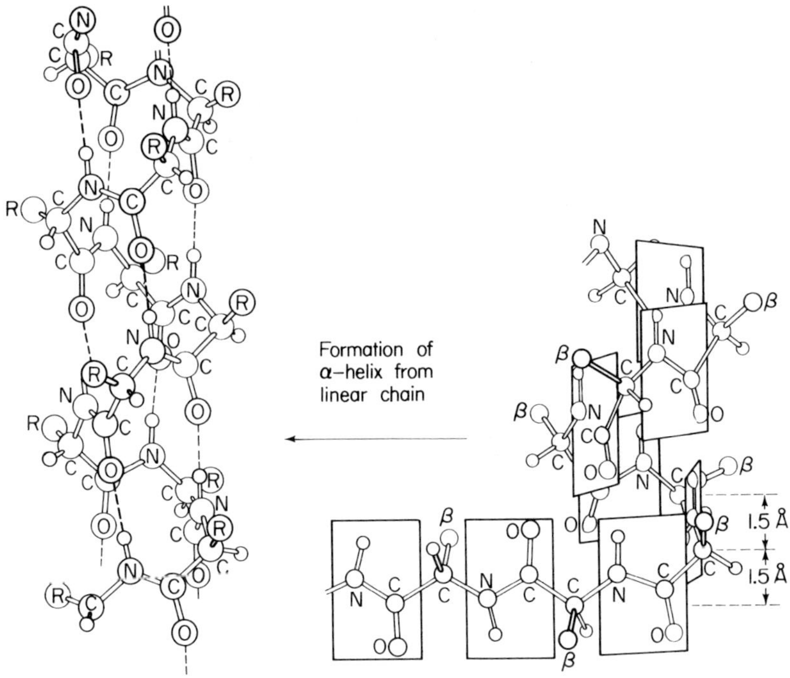

Fig. 1-6. Folding the extended polypeptide chain into a right-handed α-helix. The structure of the α-helix has all peptide bonds in intrachain hydrogen bonding. (From C. B. Anfinsen, *The Molecular Basis of Evolution*, John Wiley and Sons, New York, 1959.)

Although a hydrogen bond is a weak interaction, the very large number of possible interactions in a polypeptide chain make them an important stabilizing factor. We now consider two conformations of a polypeptide chain in which the hydrogen-bonding potential is maximal.

α-Helix

Pauling and Corey proposed the α-helical structure in 1951 as one type of secondary structure that should occur in nature because of its inherent stability (Fig. 1-6). The α-helix is a spiral arrangement of the polypeptide chain, with the chain winding around a central axis, and each amino acid residue rising along the spiral in a uniform manner. There are 3.6 amino acid residues in one complete turn of the helix, with a translational distance parallel to the helix axis of 5.4 Å (Fig. 1-7). Thus there is a rise of 1.5 Å per residue (5.4/3.6). These

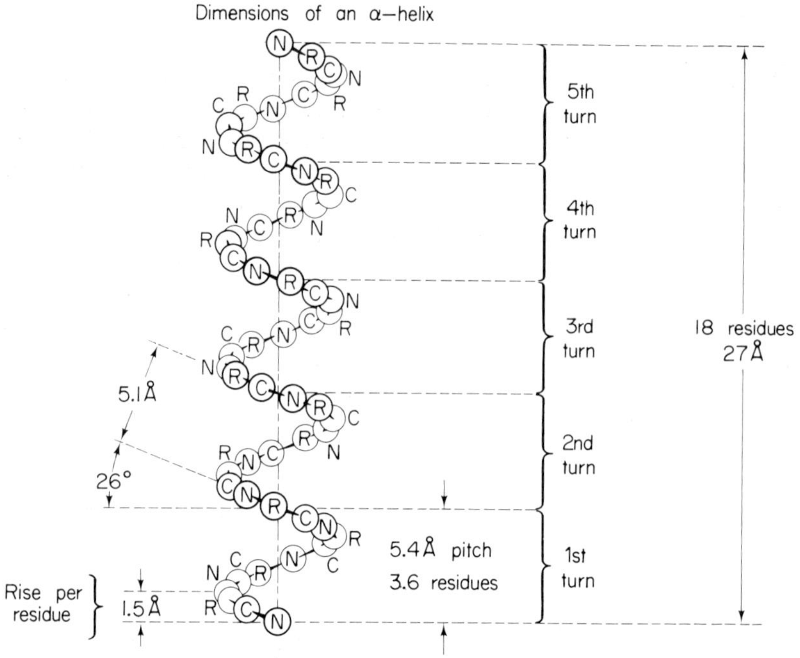

Fig. 1-7. A diagramatic view of the α-helix. The rise per residue is 1.5 Å and the translational distance is 5.4 Å per turn of 3.6 residues. (From C. B. Anfinsen, *The Molecular Basis of Evolution*, John Wiley and Sons, New York, 1959.)

dimensions are critical in order to maximize the hydrogen-bonding potential of a helical structure.

In the α-helix, every peptide carbonyl oxygen is hydrogen bonded with the amide hydrogen of the fourth amino acid residue further along the helix. These are almost ideal positions of the carbonyl and amide for the formation of a hydrogen bond, and, indeed, the hydrogen bonds lie approximately parallel to the axis of the helix. The overall effect of a helical structure and its stabilization by hydrogen bonds is to create a structure that is cylindrical in shape, fairly rigid, but capable of undergoing reversible changes in conformation on stretching and bending.

Most amino acid residues can be accommodated as structural units in the α-helix with the side chains located on the outside and pointing away from the helical axis. One of the exceptions is proline, which will not fit in a helix because of its ring structure. Those amino acids with bulky side chains, such as valine and isoleucine, also disrupt the helical structure. Amino acids that do not fit into the helix are referred to as helix "breakers."

β-Structure

A second type of secondary structure was also proposed by Pauling and Corey in 1951, an arrangement of the polypeptide chain called the β-structure, or pleated sheet. In this conformation, a number of polypeptide chains are aligned side by side, in close contact, and these chains are stabilized by intermolecular hydrogen bonding. In order to promote the maximum number of hydrogen bonds between chains, the dimensions of the polypeptide must be slightly shorter than that of a fully extended chain. As a result, the peptide backbone appears puckered and the conformation resembles a pleated sheet (Fig. 1-8). In the β-structure, the side chains are located above and below the plane of the sheet, with the direction alternating from one residue to the next.

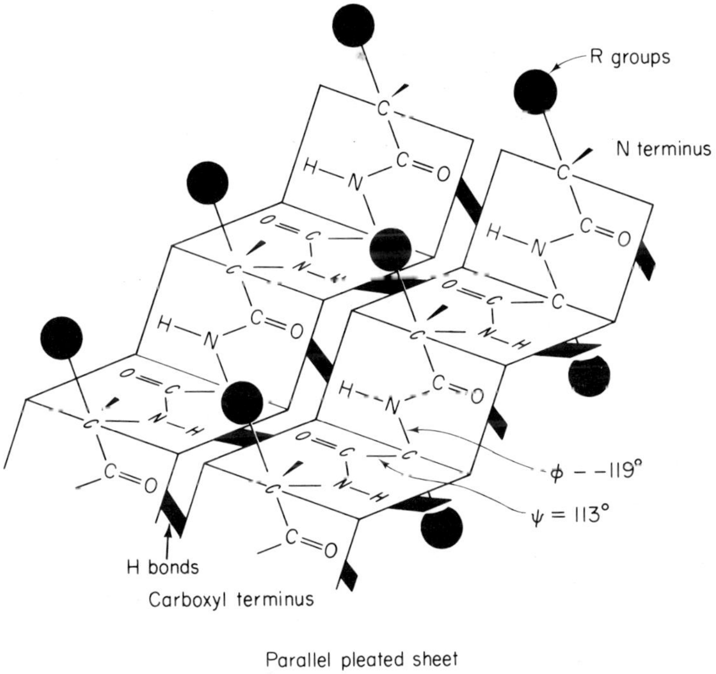

Fig. 1-8. A diagramatic view of the β-conformation or pleated-sheet structure. The polypeptide chain is oriented to facilitate intermolecular hydrogen bonding with an adjacent chain. In order to maximize the number of possible hydrogen bonds, the chains must be pleated. The side chains point away from the polypeptide backbone, alternating above and below. (From R. Barker, *Organic Chemistry of Biological Compounds*, Prentice-Hall, Englewood Cliffs, New Jersey, 1971.)

Parallel and antiparallel structures

The alignment of the peptide chains in the pleated sheet may be parallel to one another or antiparallel. In the parallel arrangement, all chains run in the same direction from the amino- to carboxyl-terminal residue, and, when antiparallel, the direction alternates from one chain to the next.

Fibrous and globular proteins

The polypeptide chains of fibrous proteins assume an α-helical structure in some cases and the β-structure in others. The α-helix is the structural feature of proteins of the α-keratin class, such as hair; the β-structure is found in the protein seracin of silk.

The fibrous proteins have a much more regular secondary structure than globular proteins. A typical globular protein cannot be described by any single type of secondary structure, but rather by a conformation that may include α-helical segments, elements of the β-structure, and regions of irregular or nonrepeating structures.

The more difficult task of elucidating the secondary structure of globular proteins was accomplished with x-ray diffraction studies of crystalline proteins. The publication of the three-dimensional electron density maps of myoglobin by Kendrew and of hemoglobin by Perutz, reported from 1961 to 1963, marked the arrival of a new era in our understanding of the detailed structure of protein molecules.

Myoglobin

The oxygen-binding protein of muscle cells, myoglobin, is a single polypeptide chain of 153 amino acids with a single heme group bound to the protein by noncovalent interactions. Kendrew established the conformation of the backbone and noted the presence of eight α-helical segments, accounting for 70 percent of the total number of residues. The α-helical segments appear as rigid cylindrical structures, with dimensions that are identical to those proposed by Pauling and Corey. The α-helical regions are connected by segments of the polypeptide chain in the extended form that permit the chain to bend and turn (Fig. 1-9). If one follows the path of the polypeptide chain starting at the amino-terminus, it is evident that the chain changes direction, crossing back and forth, and turning from front to back. The overall effect of folding is to make the structure compact and ellipsoidal in shape.

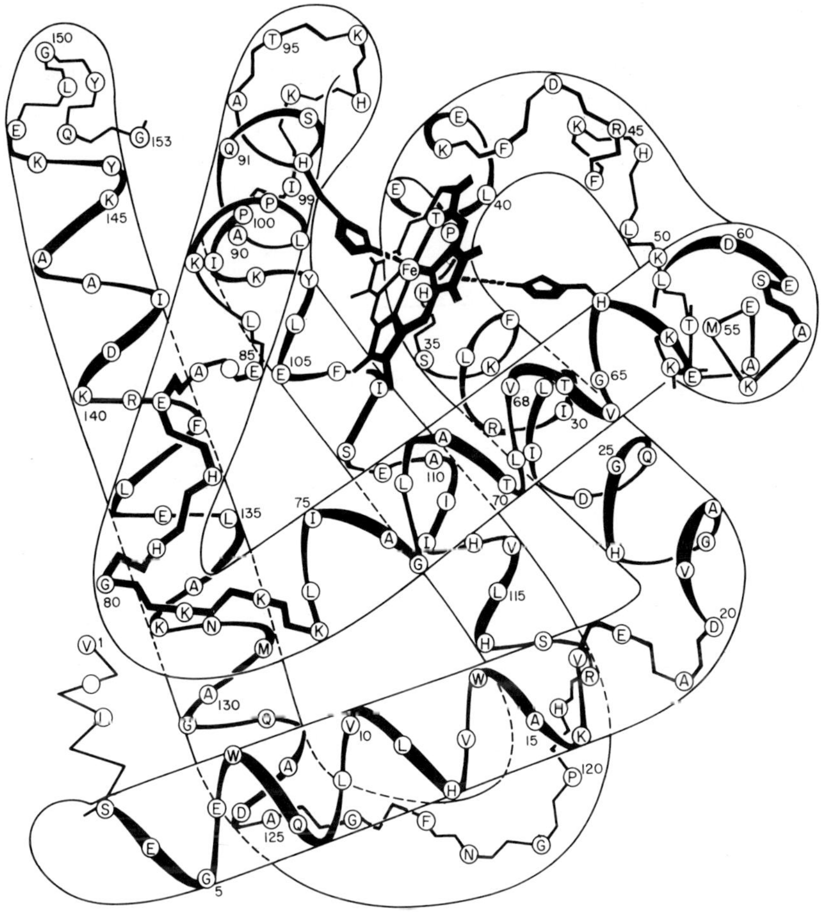

Fig. 1-9. The three-dimensional structure of the myoglobin molecule. The polypeptide chain is numbered from residue number 1, *V* for valine at the lower left, to residue 153, *G* for glycine at the upper left. The one-letter abbreviation of the amino acid residue is placed in the position of the α-carbon. The heme group is the planar structure (upper center) and α-helical sections are enclosed in the straight cylindrical segments. [M. O. Dayhoff, (Ed.), *Atlas of Protein Sequence and Structure,* Vol. **5**, National Biomedical Research Foundation, Washington, D. C., 1972.]

Other proteins

Myoglobin and hemoglobin have a higher content of α-helical regions than many other globular proteins. Recent studies on enzymes, such as hen egg-white lysozyme, bovine chymotrypsin A, bovine ribonuclease, and papain, have shown that their helical content varies from a few turns in chymotrypsin to approximately 20 percent of the total residues in papain. In the three-dimensional structure of chymotrypsin (Fig. 1-10), the polypeptide chain is extensively folded, with

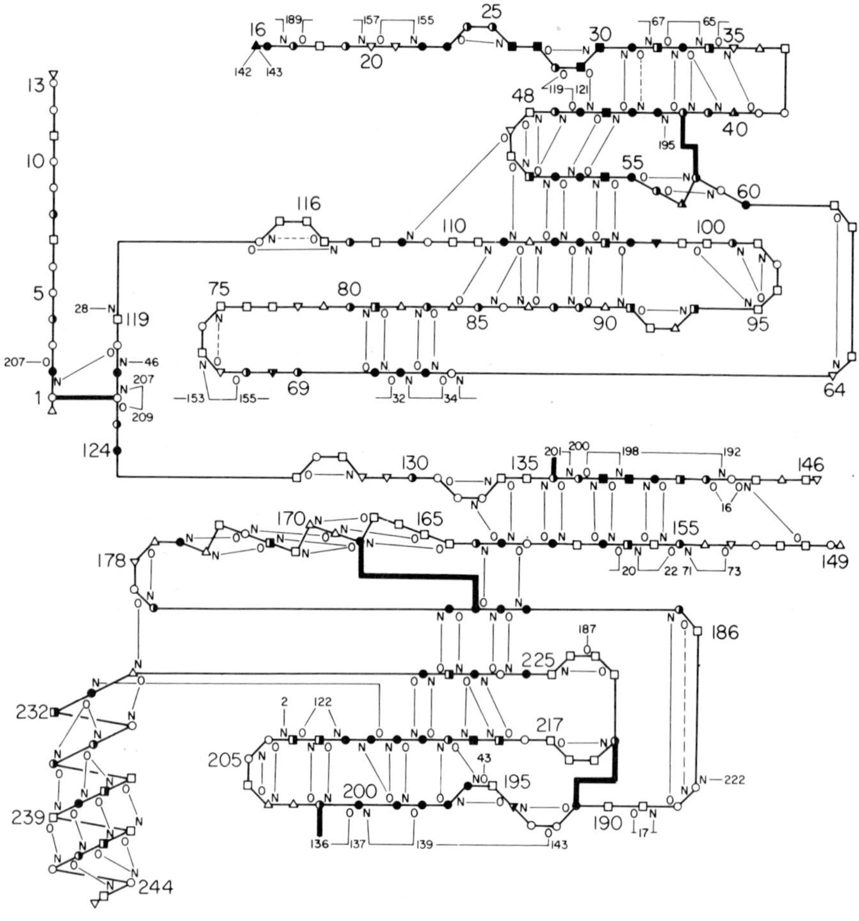

Fig. 1-10. A representation of the folding of the polypeptide chain of bovine α-chymotrypsin. Hydrogen bonds are shown as thin lines but are dashed when the bond is uncertain. Hydrogen bonding between adjacent chains is predominantly of the antiparallel pleated sheet type. Helical regions are indicated by a sawtooth pattern. The broad black bars represent disulfide bridges. Charged side chains are indicated by triangles: (▽) acidic, (△) basic, including histidine; other polar side chains by squares; and nonpolar side chains by circles. The location of each side chain in the three-dimensional structure is: filled, internal; half-filled surface; open, external. Terminal amino and carboxyl groups are similarly indicated. [From J. J. Birktoft and D. M. Blow, *J. Mol. Biol.*, **68**, 187 (1972).]

only the carboxyl-terminal eight residues in the α-helical conformation. The polypeptide chain is primarily in the extended conformation with regions of the chain running antiparallel to one another. Lysozyme and several other enzymes have regions of β-structure in addition to extended chains and α-helical sections. The β-structure is of the parallel type in lysozyme but antiparallel in ribonuclease and papain.

TERTIARY STRUCTURE OF PROTEINS

High resolution x-ray crystallography

In order to appreciate the arrangement of amino acids in globular proteins, it is necessary to examine x-ray diffraction patterns at high resolution (1.5 to 2 Å) One must look at electron-density maps representing different levels of the unit cell of the protein crystal. These three-dimensional electron densities are analyzed, in conjunction with the amino acid sequence, to determine the coordinates of the atoms of every amino acid residue. With such information available, the model builder is in a position to construct a three-dimensional structure of the protein molecule.

Rapid progress has been made on the x-ray diffraction analysis of a number of proteins that differ in biological function, size, and in the number of polypeptide chains per molecule. These studies show that a typical globular protein is compact with extensive folding of the polypeptide chain. The compact nature of the molecule places the side chains of amino acid residues in close physical contact with one another, close enough to promote side chain interactions. The tertiary structure of a protein refers to the orientation of the side chains in the folded molecule.

Examples of side chain interactions

A few representative examples of side chain interactions are illustrated in Fig. 1-11. Salt linkages and hydrogen bonding occur between polar groups; hydrophobic and van der Waals interactions occur between nonpolar groups. The disulfide bond is a type of side chain interaction of a special kind; it is a covalent bond linking together two half-cystine residues.

Although noncovalent interactions of side chains are weak interactions, the very large number of participating groups make these forces important stabilizing factors, analogous to the hydrogen bonds of the secondary structure. It should also be obvious that the tertiary and secondary structures are intimately interrelated.

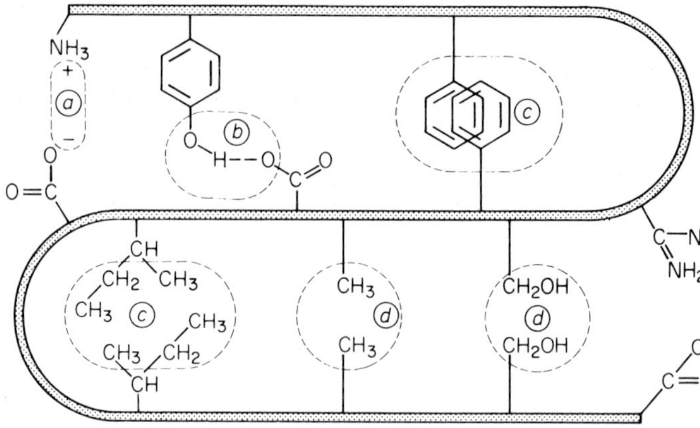

Fig. 1-11. Several types of interactions between side chain substituents which stabilize protein structure: (a) Salt linkages between ionic groups; (b) Hydrogen bonding between polar groups; (c) Hydrophobic interactions between nonpolar groups; (d) van der Waals interactions between nonpolar groups. (From C. B. Anfinsen, *The Molecular Basis of Evolution*, John Wiley and Sons, New York, 1959.)

Surface and internal features

A few rules governing the structure of proteins can be deduced from an analysis of their three-dimensional structures. X-ray studies show that globular proteins are compact because of extensive folding of the polypeptide chain. In viewing the folded structures of myoglobin, lysozyme, and ribonuclease-S, approximately half of the residues on the surface are polar and half nonpolar. Buried residues, which occur in the interior, where they are inaccessible to solvent molecules, also comprise both classes of molecules. Hydrophobic and van der Waals interactions are dominant with nonpolar side chains (Fig. 1-11).

Examples of polar and nonpolar residues

The assignment of amino acids as polar or nonpolar residues is helpful in specifying a role for the side chain groups in the structure of the protein. Based on physical-chemical studies, most notably solubility behavior in different solvents, and a knowledge of the location of residues as surface or buried, the amino acids are conveniently divided into two major categories (Table 1-2). Polar amino acids contain side chains with atoms or functional

TABLE 1-2

Classification of amino acid side chains based on physical-chemical studies and the properties of the amino acid residues deduced from their location in the three-dimensional structure of known proteins

Polar Residues	*Nonpolar Residues*	*Hybrid Types*
Serine	Glycine	Tyrosine
Threonine	Alanine	Tryptophan
Aspartic acid	Valine	
Asparagine	Leucine	
Glutamic acid	Isoleucine	
Glutamine	Proline	
Histidine	Phenylalanine	
Lysine	Tryptophan	
Arginine	Methionine	
Tyrosine	Cystine	
	Cysteine	

groups capable of forming hydrogen bonds or ionic bonds with an appropriate partner [Fig. 1-11 (a, b)]. In this group are the acidic, basic, aliphatic alcohol, and amide substituents. Polar residues are also called hydrophilic, a term that draws attention to the ability of their side chains to interact with water molecules.

Nonpolar amino acids are distinguished by the presence of aliphatic or aromatic side chain substituents. These hydrocarbonlike side chains favor self-association or hydrophobic bonding in an aqueous environment. In Figure 1-11 (c), examples are shown of those structures that self-associate when they are within van der Waals contact distance. Hydrophobic interactions make a great contribution to the conformation and stability of the native protein.

It should be noted that several amino acids contain both polar and nonpolar substitutes. Tyrosine has a polar phenolic group and a nonpolar aromatic ring, and tryptophan has a polar indole nitrogen with the remainder of the side chain nonpolar. Glutamic acid and lysine have polar functional groups but also nonpolar methylene groups. These "hybrid" types add greater flexibility to the interactions that become possible between amino acid residues in the folded state.

Buried polar groups are stabilized if they participate in hydrogen bonding or ionic bonding. The secondary structures of globular proteins contain buried regions of α-helical segments, β-structures, and hydrogen bonds between peptide bonds and side chain substituents. Buried ionic groups tend to form ion-pairs in a nonaqueous environment. The type of bonding found in a particular protein depends on its sequence, but, in general, the folded structure appears to maximize the interactions of buried polar groups.

FOLDING OF THE PROTEIN MOLECULE

Amino acid sequence dictates folding

It is evident from the preceding discussion that amino acid side chain interactions serve to stabilize the macromolecular structure in a specific conformation. It should also be recognized that these same interactions are important in the folding of a polypeptide chain from the disorganized state to the conformation typical of the biologically active molecule.

In a series of elegant experiments performed by Anfinsen and his co-workers, it was shown that the folding of a polypeptide chain is a spontaneous process, independent of cofactors and enzyme-catalyzed reactions. In the process of folding, the polypeptide chain may undergo many alterations in conformation, but it will seek that particular conformation compatible with maximum stability. An alternate way of describing the same event is to recognize that the polypeptide chain seeks a unique conformation that results in a minimization of the free energy. Thus it appears that the primary sequence of the polypeptide chain fully determines the ultimate conformation of the folded protein molecule.

These conclusions were derived from studies on the refolding of denatured ribonuclease (and other enzymes as well) where the regain of enzymic activity provided a critical test of the folding process. Ribonuclease was treated with 8 *M* urea and mercaptoethanol, the former serving to unfold the molecule and the latter causing the reduction of the four disulfide bonds. The altered molecule contained eight sulfhydryl groups; the biological activity was lost; and the physical properties agreed with those predicted for a denatured molecule (Fig. 1-12). Next, the 8 *M* urea and mercaptoethanol were removed, and the reduced protein, in an appropriate buffer, at pH 8.0, was submitted to oxidizing conditions that would cause the formation of disulfide bonds. These conditions led to the production of four disulfides with a concommittant return of enzyme activity. The properties of the recovered protein were identical with those of the native molecule. Furthermore, the yield of active molecules was high, suggesting that the equilibrium favored the folded state. (Table 1-3 shows data for ribonuclease and other proteins.)

It should be noted that there are 105 different ways to form four disulfides from eight sulfhydryl groups.[1] If folding of the polypeptide was a random process, as many as 105 different structures would be formed, with a maximum yield of only 1 percent for the correct ribonuclease structure. The high yield observed and the correct pairing of the disulfides in the regenerated ribonuclease clearly show that only one organized state is favored. Thus the amino acid sequence of

[1]Random pairing of disulfides, based on pure chance, occurs in 105 ways (8 sulfhydryls may pair in 7 ways, then 6 sulfhydryls may pair in 5 ways, 4 in 3 ways, giving a total of $7 \times 5 \times 3 \times 1$ different structures).

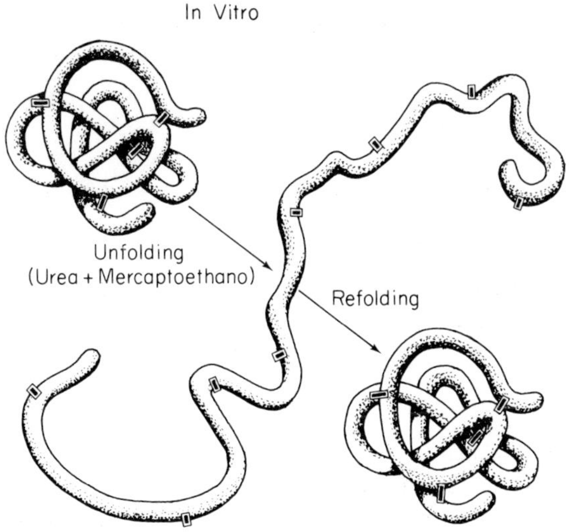

Fig. 1-12. A schematic drawing illustrating the changes in the conformation of the protein molecule in the experimental system used for studying the conversion of a linear polypeptide chain to a folded protein. A native protein, ribonuclease for example, is treated with mercaptoethanol in 8 *M* urea, which reduces the intramolecular disulfide bonds and unfolds the structure. After removal of these reagents, the reduced protein is allowed to undergo spontaneous refolding and reoxidation. [From C. J. Epstein, R. F. Goldberger, and C. B. Anfinsen, *Cold Spring Harbor Symposium*, 28, 439 (1963).]

a protein dictates the three-dimensional structure of the molecule. Presumably these experimental observations are also true of the folding process associated with the biosynthesis of protein molecules.

It is a remarkable state of affairs that protein molecules have unique structures in light of the disordered structures found with polypeptide protein fragments and with synthetically prepared "unnatural" polypeptides. It was noted by Edsall in 1968 (see reference at the end of this chapter) that the

> natural proteins, numerous as they are, represent only a tiny fraction of the vast number of possible combinations of amino acid residues.[2] The great majority of these possible sequences may have no tendency, in any solvent medium whatever, to assume a unique conformation. Natural selection would in general eliminate such sequences, if they arose in an organism by mutation, as biologically useless [see Chap. 11].

[2]For a protein with 100 residues, about 10^{130} different structures can be formed from a mixture of 20 different amino acids.

TABLE 1-3

"Refolding" of Disulfide-Bond Containing Proteins (following Reduction and Unfolding)[a]

Protein	*Disulfide Bonds*	*Recovery of Activity (percent)*
Ribonuclease A (bovine pancreatic)	4	95–100
Lysozyme (egg white)	4	50–80
Taka-amylase A (Aspergillus oryzae)	4	48
Trypsin (bovine)	6	
CM-cellulose trypsin		4
Poly-DL-alanyl-trypsin		8
Proinsulin (bovine and rat)	3	40–80
Insulin (bovine)	3	5–10
Alkaline phosphatase (E. coli)	2	80
Pepsinogen (swine)	3	50
Serum albumin (human and bovine)	17	50
Ovomucoid (turkey)	8	90
Trypsin inhibitor (soybean)	2	40

[a]Adapted from C. J. Epstein, R. F. Goldberger, and C. B. Anfinsen, *Cold Spring Harbor Symposium*, **28**, 439 (1963).

REFERENCES

Books

S. B. Needleman (Ed.), *Protein Sequence Determination*. Springer-Verlag, New York, 1970.

H. D. Law, *The Organic Chemistry of Peptides*. Wiley-Interscience, New York, 1970.

R. E. Dickerson and I. Geis, *The Structure and Action of Proteins*. Harper and Row, New York, 1969.

W. A. Schroeder, *The Primary Structure of Proteins*. Harper and Row, New York, 1968.

D. T. Elmore, *Peptides and Proteins*. Cambridge University Press, New York, 1968.

J. Liggett-Bailey, *Techniques in Protein Chemistry*, 2nd ed. Elsevier, Amsterdam, 1967.

Reviews

C. C. F. Blake, "X-Ray Studies of Crystalline Proteins," *Progress in Biophysics,* **25**, 83 (1972).

F. M. Richards, H. W. Wyckoff, W. D. Carlson, N. M. Allewell, B. Lee, and Y. Mitsui, "Protein Structure, Ribonuclease-S and Nucleotide Interactions," *Cold Spring Harbor Symposia on Quantitative Biology*, **36**, 35 (1972).

J. T. Edsall, "Thoughts on the Conformation of Proteins in Solution," in *Structural Chemistry and Molecular Biology*, A. Rich and H. Davidson (Eds.). W. H. Freeman & Co., San Francisco, 1968, p. 88.

R. E. Dickerson, "X-Ray Analysis and Protein Structure," *The Proteins*, **2**, 603 (1964).

J. A. Schellman and C. Schellman, "The Conformation of Polypeptide Chains in Proteins," *The Proteins*, 2, 1 (1964).

C. J. Epstein, R. F. Goldberger, and C. B. Anfinsen, "The Genetic Control of Tertiary Structure: Studies with Model Systems," *Cold Spring Harbor Symposia on Quantitative Biology*, 28, 439 (1963).

H. K. Schachman, "Considerations on the Tertiary Structure of Proteins," *Cold Spring Harbor Symposia on Quantitative Biology*, 28, 409 (1963).

J. T. Edsall and J. Wyman, *Biophysical Chemistry*, Vol. 1. Academic Press, New York, 1959, Chapter 3, "Problems of Protein Structure."

C. B. Anfinsen, "Principles that Govern the Folding of Protein Chains," *Science*, **181**, 223 (1973).

S. Moore and W. H. Stein, "Chemical Structures of Pancreatic Ribonuclease and Deoxyribonuclease," *Science*, **180**, 458 (1973).

Purification of Proteins

TWO

In this chapter we review the principal procedures commonly used for the purification of proteins, beginning with the problem of choosing an appropriate biological sample and concluding with a few comments on those methods that are used to establish the homogeneity of the isolated sample. The number and type of steps encountered between the beginning and the end are dependent on the initial concentration of the desired protein in the starting sample and on the nature of the other components present. Indeed, it is impossible to describe any single purification scheme that would be ideal for the isolation of more than one protein. Instead a few basic concepts and several empirical considerations can be offered in the hope that they will be helpful in the solution of a specific problem.

Since proteins are macromolecules and their biological properties are dependent on a specific three-dimensional structure, many of the purification procedures that are effective for small molecules are unsuitable for proteins and would cause their denaturation. Those physical forces and chemical agents that are capable of breaking the noncovalent interactions that stabilize the protein molecule must be avoided or kept to a minimum. As a general rule, elevated temperatures, high or low pH, organic solvents, and detergents represent conditions and agents that may denature a protein molecule. Unfortunately, susceptibility to different conditions varies among proteins, and preliminary experiments must be performed in order to test new procedures. One or more physical properties of a protein could be measured to judge stability, but such tests are generally less satisfactory than the direct test of the biological activity of the molecule.

ANALYTICAL TECHNIQUES

Measurement of enzyme activity

The biological activity associated with a protein molecule is a highly specific property, and assays of such activity can be extremely sensitive, usually require only micrograms of a sample, and are easily performed using chemical reactions or the response of tissues or of living organisms. It is probable that the simplest biological activity to measure is that of an enzyme. Since an enzyme is a biological catalyst, its activity is determined by measuring the change it produces in the rate of a chemical reaction. The rate of an enzyme-catalyzed reaction is estimated from changes in the utilization of reactants or in the appearance of products. The rate can be expressed as the change in the concentration of either substrate or product per unit of time, at a specified temperature; and the units may be in terms of a micromolar amount, a percentage change, or in arbitrary units–for example, an absorbancy value–and time is usually in minutes or seconds. The specific activity of an enzyme-catalyzed reaction is defined as the rate of the reaction per unit of protein, with the protein concentration expressed in micromoles, in milligrams, or in arbitrary units, such as absorbancy at a wavelength of 280 nm. If the protein concentration is expressed as the molarity of active enzyme molecules, the specific activity calculation is then referred to as the catalytic rate constant or turnover number.

Measurement of other proteins

The activity of a protein hormone is estimated from the changes it produces in the metabolism of the whole animal or of isolated tissues. Structural proteins or proteins lacking a distinctive biological activity are best examined by chromatographic procedures, by electrophoretic and ultracentrifugal analysis, or by immunological techniques. Proteins that contain firmly bound prosthetic groups, cofactors, or metal ions may be estimated from the content of the nonprotein constituent.

SOURCE OF PROTEINS

The selection of a biological source for the isolation of a specific protein is dictated by the interests of the investigator in animal, plant, or bacterial systems and by the relative abundance of the protein in different species. There is an obvious advantage in utilizing a source in which the protein is present in large amounts.

The study of mammalian proteins is dictated by the availability of the starting material; slaughterhouse tissues and organs of the cow, sheep, pig, and horse, laboratory animals, human blood, urine, and autopsy material are representative of several of the most common sources. Plant systems are usually restricted to the study of the proteins of the leaf, fruit, and latex. Microbiological systems are used extensively, both as wild type and as mutant strains.

EXTRACTION

The cellular organization of a biological system is disrupted by mincing, grinding, or homogenizing the sample, and the cell contents are extracted with an appropriate solvent. Water is suitable for proteins of the albumin class, and the stability and solubility may be increased if the pH of the extract is controlled with an appropriate buffer. Proteins of the globulin type show increased solubility in the presence of dilute salt solutions; 0.1 *N* NaCl is typical. Proteolipids, for example, are extracted with dilute solutions of detergents, such as sodium dodecyl sulfate, digitonin, or bile salts, or with organic solvents. Often other water-insoluble proteins may be brought into solution with dilute alkali, although the treatment may be capable of partly altering the structure.

All too often the disruption of the organized state of a cell releases proteolytic enzymes into the medium. The proteins of the cell may then be subject to degradation by these proteolytic enzymes, resulting in a decrease in the amount of native protein. Autolysis reactions may be inhibited or slowed down by adding diisopropylphosphofluoridate as a group specific inhibitor of serine proteinases and iodoacetate for the inhibition of those proteolytic enzymes that require sulfhydryl groups at their active sites (Chap. 11). These substances inhibit enzymes by reacting irreversibly with components of the active site. Autolysis can also be decreased by choosing a pH, temperature, or organic solvent suitable for the extraction of the desired protein but unfavorable to the proteolytic enzymes.

In addition to the loss of biologically active proteins, the action of proteolytic enzymes is also capable of creating altered active molecules by partial degradation of a precursor molecule. A notable example is bovine carboxypeptidase A, an enzyme isolated from bovine pancreas in more than one form, where the molecules vary in the number of residues at the amino-terminal end.

Proteolytic enzymes in the extraction mixture are also capable of converting an inactive precursor molecule to the active species. For example, the zymogen molecules of the pancreas are usually modified by partial proteolytic digestion to form the enzymes trypsin, chymotrypsin, and carboxypeptidase. Activation may occur during the extraction and subsequent purification. The conversion of a zymogen to an active protein may proceed so quickly that the precursor may be overlooked; this is what has happened in the case of proinsulin, the inactive precursor of insulin (Chap. 1). In 1967 proinsulin was discovered and its structure was shown to be related to that of insulin.

In the preparation of an extract of a biological sample, a large volume of extracting fluid per weight of starting sample is usually used in order to increase the yield of the desired protein. Before proceeding to further fractionation, it is often convenient to reduce the volume of the extract, which may be several liters or more, to a smaller volume more suitable to normal laboratory operations. High concentrations of a neutral inorganic salt, such as saturated solutions of ammonium sulfate, sodium chloride, or sodium phosphate, which reduce the solubility of aqueous solutions of proteins to a very low level, may be added to precipitate the protein. Frequently, the protein mixture is recovered as an insoluble fraction in essentially a quantitative yield.

FRACTIONATION METHODS

Selective precipitation

One approach to the fractionation of a very impure protein sample is the use of selective precipitation. If a neutral salt or organic solvent is added in graded amounts, the least-soluble proteins are precipitated first, followed sequentially by proteins of increasing solubility. Usually fractionation of a protein mixture is performed by increasing the concentration of a neutral salt to a level at which unwanted proteins are first precipitated. After removal of protein impurities, salt is added again to precipitate a fraction containing the protein of interest. If the protein is very soluble in water, the fractionation may be performed at the isoelectric point of the protein as an effective way to decrease its solubility. The isoelectric pH is a distinctive property of a protein, so that the use of pH control alone is also a useful procedure for selective precipitation.

The solubility of proteins varies with ionic strength, pH, temperature, and the polarity of the solvent. The classical studies on the solubility of horse carboxyhemoglobin as a function of the concentration of a neutral salt are summarized graphically in Fig. 2-1. At low concentrations of salt, where the ionic strength is small, the solubility of the protein is enhanced—a situation described as a salting-in effect. However, as more salt is added and the ionic strength increases, the solubility of the protein continually falls off, leading to precipitation or salting-out. This behavior is typical of water-soluble proteins.

pH variation

The influence of pH on solubility is seen in the studies with β-lactoglobulin at different concentrations of a neutral salt (Fig. 2-2). Minimum solubility is observed in the isoelectric region of the protein with solubility increasing on either side of the isoelectric point.

The polarity of the solvent also influences solubility. Solubility decreases on addition of organic solvents miscible with water; typical solvents are ethanol and acetone. Care must be exercised in the use of organic solvents since they are

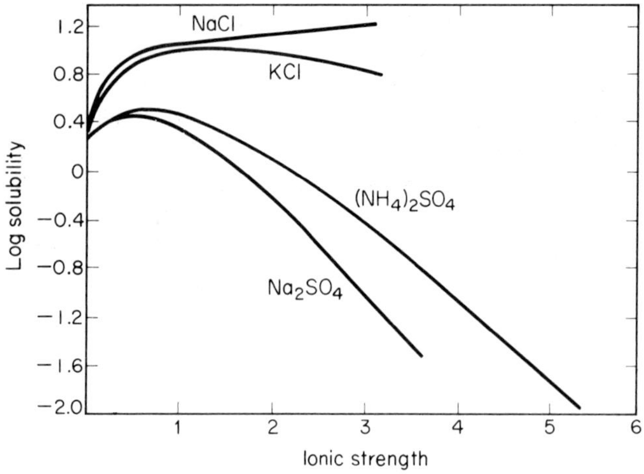

Fig. 2-1. The solubility of horse carboxyhemoglobin in salt solutions of varying ionic strength. (From J. T. Edsall and J. Wyman, *Biophysical Chemistry*, Vol. 1, Academic Press Inc., New York, 1958.)

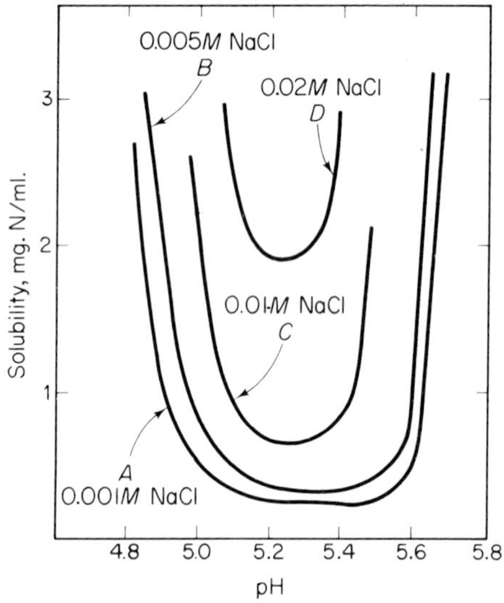

Fig. 2-2. The solubility of β-lactoglobulin as a function of pH at four different concentrations of sodium chloride. (From A. White, P. Handler, and E. L. Smith, *Principles of Biochemistry*, 4th Edition, McGraw-Hill, New York, 1968.)

generally good protein denaturing agents. Proteins are more stable at low temperatures, and thus precipitation with organic solvents is generally performed near the freezing point of the water-solvent mixture.

COLUMN CHROMATOGRAPHY

Further purification of the protein mixture is normally accomplished with one or more chromatographic procedures. A chromatographic system is almost always incorporated into a purification scheme, for the technique has a high resolving power toward similar compounds. The resolving power of the system is based on the differential distribution of components between two phases, which, in column chromatography, may consist of a liquid and a solid phase. Although two components may resemble one another in many structural features, they can be resolved if the distribution process is repeated again and again. Chromatographic systems are capable of separating components that differ slightly in their distribution coefficients if a sufficient number of distributions are made.

In column chromatography, the solid phase is the chromatographic medium and it is contained in a long glass column. The liquid phase is usually a buffer, which is allowed to percolate through the column at a controlled flow rate. A very large number of distributions is made in a continuous process as the liquid phase travels through the column. The resolution is proportional to the height of the column packing, and the sample capacity is related to the diameter of the column.

Ion-exchange chromatography

Column chromatography of proteins on ion-exchange resins separates components based on differences in their charge and, to a lesser extent, on adsorption. Ion-exchange resins are high molecular weight polymers with functional groups that are acidic (cation exchangers) or basic (anion exchangers), while the polymeric matrix itself is stable to organic solvents and to solutions of both high and low pH. Suitable polymeric matrices include polyvinyl styrene cross-linked with divinyl benzene, cellulose, or high molecular weight dextrans. Table 2-1 lists several of the acidic and basic groups commonly used in commercially available ion exchangers.

The equilibria existing between the functional groups of the resin and protein are depicted in Eq. (2-1) for a cation exchanger. The ion exchanger in this

$$\text{Resin-COO}^-\text{Na}^+ + {}^+\text{NH}_3\text{-Protein} \rightleftharpoons \text{Resin-COO}^-\,{}^+\text{NH}_3\text{-Protein} + \text{Na}^+ \quad (2\text{-}1)$$

example is of the carboxyl type with sodium ions acting as counter-ions to establish electrical neutrality. The available cationic groups of the protein, represented as NH_3^+-protein, displace sodium ions and bind to the resin by electrostatic interactions. The ion-exchange equilibria are influenced by pH and the ionic strength of the buffer. At low pH values, binding of the protein to the resin is favored, whereas high pH values or an increase in the ionic strength displaces the protein from its salt linkage with the carboxylate groups.

TABLE 2-1

Ion-Exchange Adsorbents Suitable for Protein Purification[a]

Anion Exchangers	*Ionizable Group*	*Matrix Available*
Aminoethyl- (AE-)	$-O-CH_2-CH_2-NH_2$	Fibrous cellulose
Diethylaminoethyl- (DEAE-)	$-O-CH_2CH_2-N(CH_2CH_3)_2$	Fibrous cellulose, micro-granular cellulose, Sephadex, polyacrylamide gel
Cation Exchangers	*Ionizable Group*	*Matrix Available*
Carboxymethyl- (CM-)	$-O-CH_2-COOH$	Fibrous cellulose, micro-granular cellulose, Sephadex, polyacrylamide gel
Phospho- (P-)	$-O-P(=O)(OH)-OH$	Fibrous cellulose, Sephadex
Sulfoethyl- (SE-)	$-O-CH_2-CH_2-S(=O)_2-OH$	Fibrous cellulose Sephadex

[a]Adapted from S. R. Himmelhoch, *Meth. in Enzymology,* **22**, 273 (1971).

Proteins are polyvalent molecules and at a given pH, different proteins have different numbers and types of charged groups that can interact with the ion-exchange resin. It is this tremendous variety of possible interactions between protein and the ion exchanger that leads to the high resolving power of the technique.

An example of the chromatography of pancreatic ribonuclease on a carboxylic resin at a fixed pH and ionic strength illustrates the application of the chromatographic procedure for the final purification of the enzyme (Fig. 2-3). The ribonuclease fraction (peak A) has the properties of a homogeneous sample, and, indeed, the amino acid sequence of the protein was determined with chromatographically purified samples. The purification of swine kidney leucine aminopeptidase was accomplished by displacement from a cellulose anion exchanger with a linear gradient of gradually increasing ionic strength (Fig. 2-4).

A difficult purification problem may frequently require the use of more than one type of ion-exchange procedure. It is possible that the first may use a cation exchanger followed by a second column separation on an anion exchanger, or with the reverse order (Fig. 2-5). The widespread adoption of ion-exchange chromatography has contributed to the successful purification of many different proteins.

In addition to the ion-exchange process, partition and absorption chromatography have occasionally been used for the fractionation of protein mixtures.

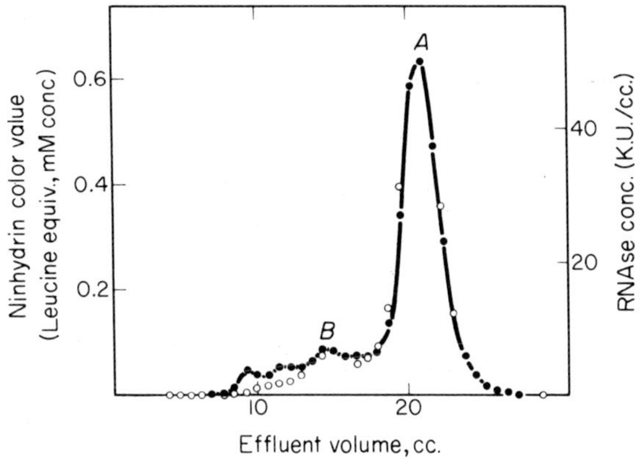

Fig. 2-3. Chromatography of pancreatic ribonuclease on the carboxylic ion-exchange resin, IRC-50 (XE-64). Column dimension, 0.9 × 30 cm; eluant, 0.2 *M* sodium phosphate buffer, pH 6.47. Ninhydrin value, •; ribonuclease activity, ○. [From C. H. W. Hirs, S. Moore, and W. H. Stein, *J. Biol. Chem.*, **200**, 493 (1953).]

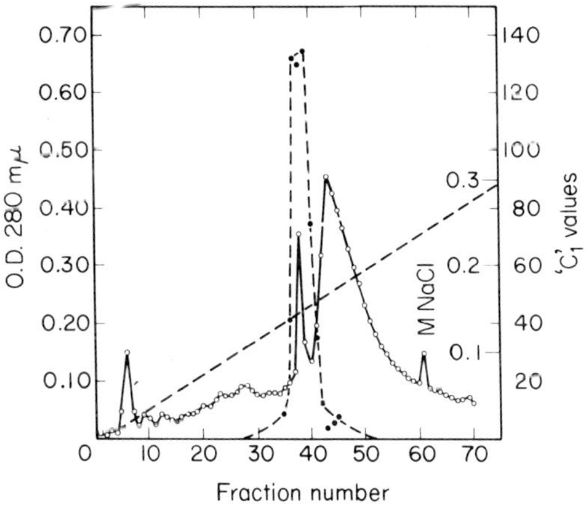

Fig. 2-4. Chromatography of swine kidney leucine aminopeptidase on the anion exchanger, DEAE-cellulose. Column dimension, 0.9 × 25 cm; eluant, 0.005 *M* Tris-HCl buffer, pH 8.0, containing 0.005 *M* $MgCl_2$, employing a linear gradient from 0 to 0.3 *M* NaCl. Absorbance at 280 nm, ○; enzyme activity, •. [From J. E. Folk, J. A. Gladner, and T. Viswanatha, *Biochem. Biophys. Acta,* **36**, 256 (1959).]

Separations based on differences in charge can also be accomplished by electrophoresis on a supporting medium of cellulose, starch, or polyacrylamide gels (see later).

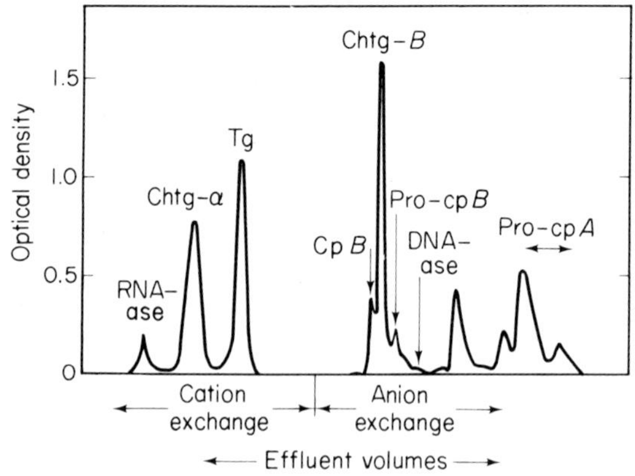

Fig. 2-5. Composite drawing summarizing the results of two chromatograms of bovine pancreatic juice. A cellulose ion-exchange column (DEAE-cellulose) was used to separate eight anionic protein components. The cationic components were resolved on IRC-50 (XE-64). An initial separation of the pancreatic proteins into anionic and cationic components was made on DEAE-cellulose. The abbreviations are RNA-ase, ribonuclease; ChTg-α, chymotrypsinogen A; Tg, trypsinogen; Pro-Cp B and Cp-B, procarboxypeptidase B and carboxypeptidase B; DNA-ase, deoxyribonuclease; Pro-Cp A, procarboxypeptidase A. [From P. J. Keller, E. Cohen, and H. Neurath, *J. Biol. Chem.*, **233**, 344 (1958).]

Gel filtration

Proteins differing in size may be separated by a chromatographic procedure, which has been termed gel filtration, in which beads of cross-linked dextran gels or polyacrylamide gels serve as the matrix (see Chap. 3). These high molecular weight, insoluble polymers are produced commercially with different degrees of cross-linkage, each capable of allowing molecules below a specified size to penetrate their interstices but excluding all higher molecular weight molecules. The order of elution of a mixture of proteins would be the highest molecular weight molecules first, followed by those of intermediate size, and with the smallest molecular weight components last. Since the molecular weight range of known proteins varies from several thousand to several million, gel filtration is a valuable step in purification, capable of resolving mixtures of components that show relatively small size differences (Chap. 3).

Gel filtration procedures were incorporated into the purification scheme developed for staphylococcal nuclease. In a typical experiment the relatively low molecular weight enzyme (17,000) was partially resolved from larger protein contaminants (Fig. 2-6) by passage through a column of Sephadex G-75 (see Chap. 3).

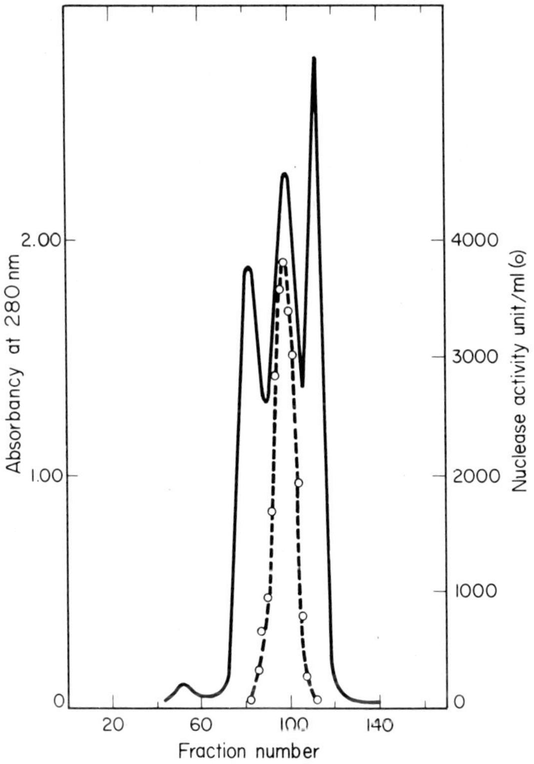

Fig. 2-6. Gel filtration of staphylococcal nuclease on a column (3 × 250 cm) of Sephadex G-75 (fine). The column was equilibrated and operated with a 0.3 *M* ammonium acetate buffer at pH 6.0. Fractions of approximately 18 ml were collected. Protein content was determined spectrophotometrically, and enzyme activity was estimated from the rate of hydrolysis of DNA. (From unpublished results of J. L. Bohnert and H. Taniuchi, National Institutes of Health, Bethesda, Md.)

Affinity chromatography

In most of the procedures described above, purification is accomplished with methods that take advantage of differences in solubility, charge, size, or absorptive properties. A different approach has been developed for the specific binding of a biologically active protein to a solid support. Advantage is taken of the known specificity of the protein for a small molecule, such as a substrate, inhibitor, or activator. For example, a competitive inhibitor of an enzyme may be covalently attached to an insoluble polymer. On addition of a tissue extract, or a partially purified sample, only those proteins with an affinity for the attached ligand will interact and form a stable complex (Fig. 2-7). Impurities are removed on further washing, and the desired protein is later recovered by displacement with a high concentration of a salt solution, with alterations of pH, or, most specifically, with a high concentration of a competitive ligand. Since purification depends on a specific interaction of the protein with a complementary structure (the ligand) and since chromatographic techniques are commonly used for the separation, the name "affinity chromatography" is reserved for this procedure.

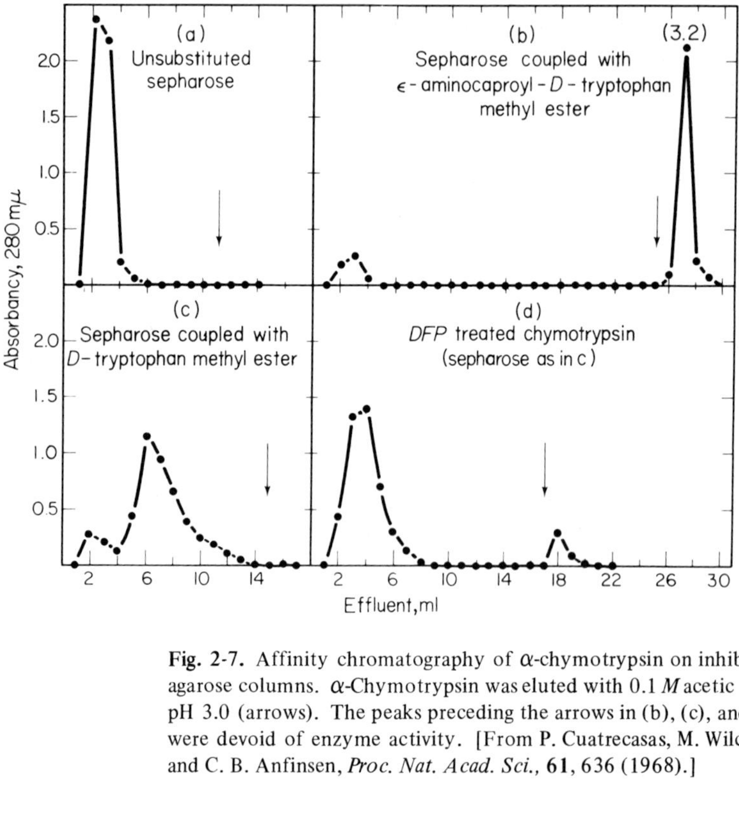

Fig. 2-7. Affinity chromatography of α-chymotrypsin on inhibitor-agarose columns. α-Chymotrypsin was eluted with 0.1 *M* acetic acid, pH 3.0 (arrows). The peaks preceding the arrows in (b), (c), and (d) were devoid of enzyme activity. [From P. Cuatrecasas, M. Wilchek, and C. B. Anfinsen, *Proc. Nat. Acad. Sci.*, **61**, 636 (1968).]

Several insoluble polymers are suitable for affinity chromatography, such as aminoethyl cellulose, polyacrylamide beads, and cross-linked dextran gels. Better chromatographic conditions are found with agarose, an insoluble polysaccharide gel, and details of its use for the purification of enzymes will be emphasized. Porath showed that cyanogen bromide forms an "activated" compound, an iminocarbonate, with vicinal hydroxyl groups of the dextran gel [Eq. (2-2)].

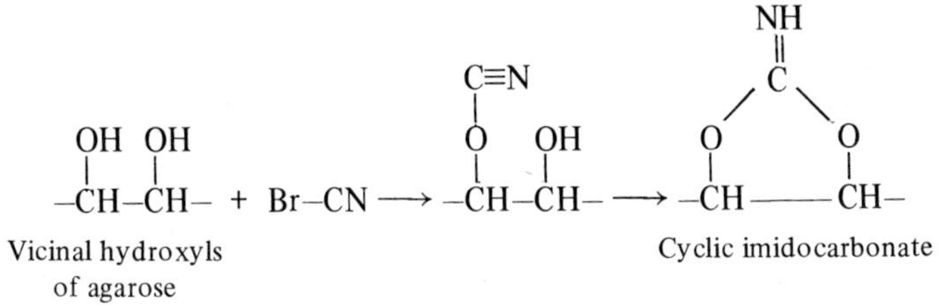

(2-2)

Further treatment of the iminocarbonate derivative with an amino compound provides a simple way to attach different structures to the agarose structure [Eq. (2-3)].

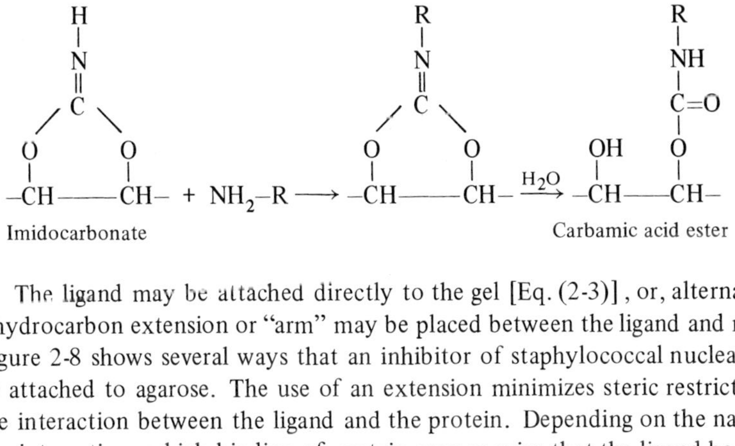

(2-3)

The ligand may be attached directly to the gel [Eq. (2-3)], or, alternatively, a hydrocarbon extension or "arm" may be placed between the ligand and matrix. Figure 2-8 shows several ways that an inhibitor of staphylococcal nuclease may be attached to agarose. The use of an extension minimizes steric restrictions in the interaction between the ligand and the protein. Depending on the nature of the interaction, a high binding of protein may require that the ligand be placed at a considerable distance from the matrix.

The major advantages of affinity chromatography are speed and specificity, with the possibility that purification may be accomplished as a single-step procedure. The versatility of the method is increased by the use of other relatively specific ligands and new procedures for different types of proteins: organic mercurials for the separation of thiol proteins; the immobilation of protein inhibitors or antibodies specific for enzymes or antigens; and the use of specific chemical cleavage of the ligand-protein complex to remove bound protein. Affinity chromatography has been used to purify enzymes (Fig. 2-9), antibodies, receptor sites of steroids, and chemically synthesized polypeptides. Further applications are limited only by the imagination of the investigator and the properties of the individual protein.

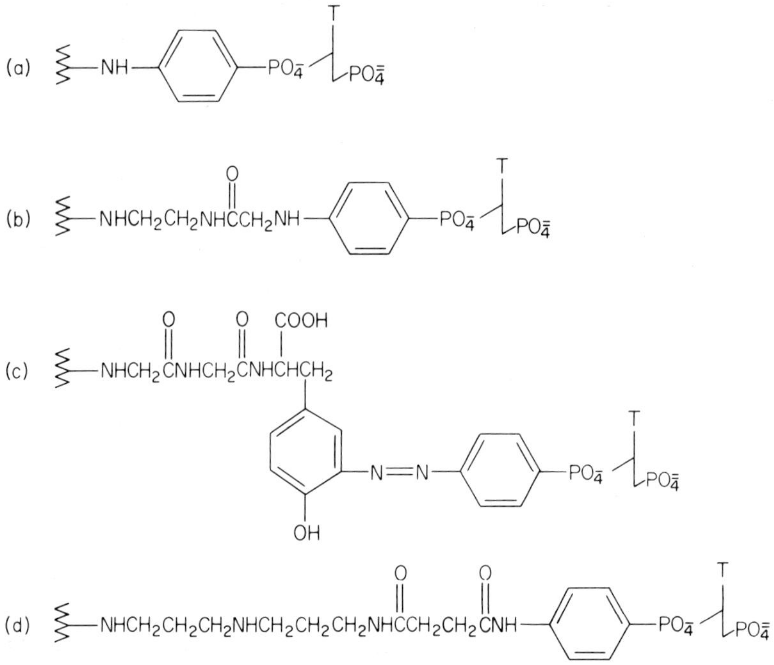

Fig. 2-8. Specific staphylococcal nuclease affinity chromatographic adsorbents prepared by attaching the competitive inhibitor, pdTp-aminophenyl, to various derivatives of Sepharose-4B or Bio-Gel P-300. In (a), the inhibitor was attached directly to agarose, after activation of the gel with cyanogen bromide, or to polyacrylamide via the acyl azide step. In (b), ethylenediamine was reacted with cyanogen bromide-activated Sepharose, or with the acyl azide polyacrylamide derivative. This amino gel derivative was then reacted with *N*-hydroxysuccinimide ester of bromoacetic acid to form the bromoacetyl derivative; the latter was then treated with the inhibitor. In (c), the tripeptide, Gly-Gly-Tyr, was attached by the α-amino group to agarose or polyacrylamide by the cyanogen bromide or acyl azide procedure, respectively; this gel was then reacted with the diazonium derivative of the inhibitor. In (d), 3,3′-diaminodipropylamine was attached to the gel matrix by the cyanogen bromide or acyl azide step. The succinyl derivative, obtained after treating the gel with succinic anhydride in aqueous media, was then coupled with the inhibitor with a water-soluble carbodiimide. The jagged vertical lines represent the agarose or polyacrylamide backbone. [From P. Cuatrecases, *J. Biol. Chem.*, **245**, 3059 (1970).]

CRYSTALLIZATION

Before chromatographic methods were available and solubility differences were primarily exploited, the crystallization of a protein was considered essential

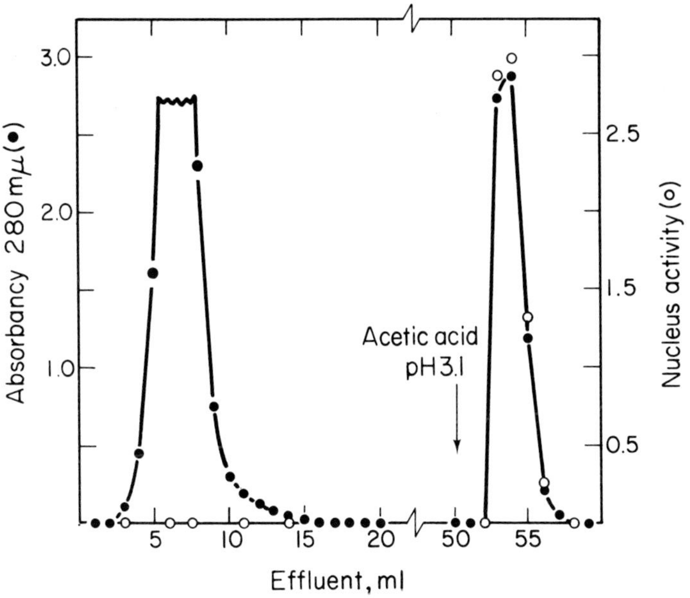

Fig. 2-9. Purification of staphylococcal nuclease by affinity chromatography on a nuclease-specific Sepharose column [see Fig. 2-8(a)]. The column was equilibrated with 0.05 *M* borate buffer, pH 8.0, containing 0.01 *M* $CaCl_2$. Approximately 40 mg of partially purified nuclease was applied in the buffer. At the position of the arrow, elution with buffer was stopped and 0.1 *M* acetic acid was added to elute nuclease. [From P. Cuatrecasas, M. Wilchek, and C. B. Anfinsen, *Proc. Nat. Acad. Sci.*, **61**, 636 (1968).]

as a purification step if a sample was to be brought to a state of homogeneity. It is now quite clear that a crystalline sample need not be pure, although repeated crystallization can sometimes be useful to improve the purity of a sample. Current interest in the crystallization of proteins is mainly directed toward the preparation of samples suitable for x-ray diffraction studies. Crystals used for x-ray studies are of large size, approximately 0.1 nm or greater, and appropriate conditions must be found for the nucleation and growth of crystals to reach this size.

CRITERIA OF PURITY

The task of establishing the purity of a protein sample is far from simple and subject to many limitations. The high molecular weight of a typical protein makes it difficult to detect small amounts of contaminating proteins, since the molar amount present may be below the limits of detection of most analytical procedures. It is possible, for example, for two or more proteins to behave in a similar fashion when the sample is examined with any single test of purity. This behavior may create the false impression that the sample is homogeneous when it is really a mixture. The difficulty in separating components necessitates the selection of more than one test of purity; only if the sample passes all tests can one consider it to be homogeneous. Indeed, it must be recognized that a test of purity can never show that a sample is pure; rather, it only offers evidence that another component is absent. Consequently, methods of high resolving power must be used in conjunction with sensitive detection devices for specific properties of the protein or of suspected contaminants.

One such specific property of a protein is its biological activity and this serves as an excellent test of purity. The biological activity, expressed in specific activity units, is an index of purity because the specific activity increases as contaminants are removed and it reaches a constant and maximal value when the sample cannot be purified further. Furthermore, the disappearance of the biological activity of a contaminant is an indirect assessment of purity. If the protein contains an easily measurable metal or tightly attached cofactor, the content of the component should increase on purification and become constant when the sample reaches a state of homogeneity. Immunological techniques are also useful in judgments of purity.

Additional tests of purity usually involve the use of methods that measure hydrodynamic properties of proteins. The ultracentrifuge is capable of separating components based on their size and shape (Chap. 3). A symmetrical distribution of a component is indicative of a homogeneous sample.

Electrophoresis of a mixture separates components that differ in their electrical charge. Several pH values are generally tried to vary the charge on the components. Electrophoretic studies are carried out either with the sample free in solution with the use of a Tiselius cell or with a solid support as the electrophoretic medium. In the latter category, disc gel electrophoresis, in particular, displays a high resolving power and a great sensitivity and serves as a technique that is rapidly replacing other electrophoretic procedures.

Disc gel electrophoresis

Disc gel electrophoresis has become a major item in the arsenal of analytical methods employed by the protein chemist, because of its high resolving power and versatility, the small amounts of sample required, and the simplicity of the equipment needed. From a schematic diagram of a typical unit, Fig. 2-10, it is clear that the unit may be assembled easily and with minimum expense.

The electrophoretic separation is performed in a tube containing a separating gel in the lower part and a concentrating gel (stacking gel) in the upper part. A sample is applied to the upper gel surface and the components are resolved in the main part of the separating gel on application of an electrical potential. At the completion of the electrophoretic separation, the tubes are removed, the gels are

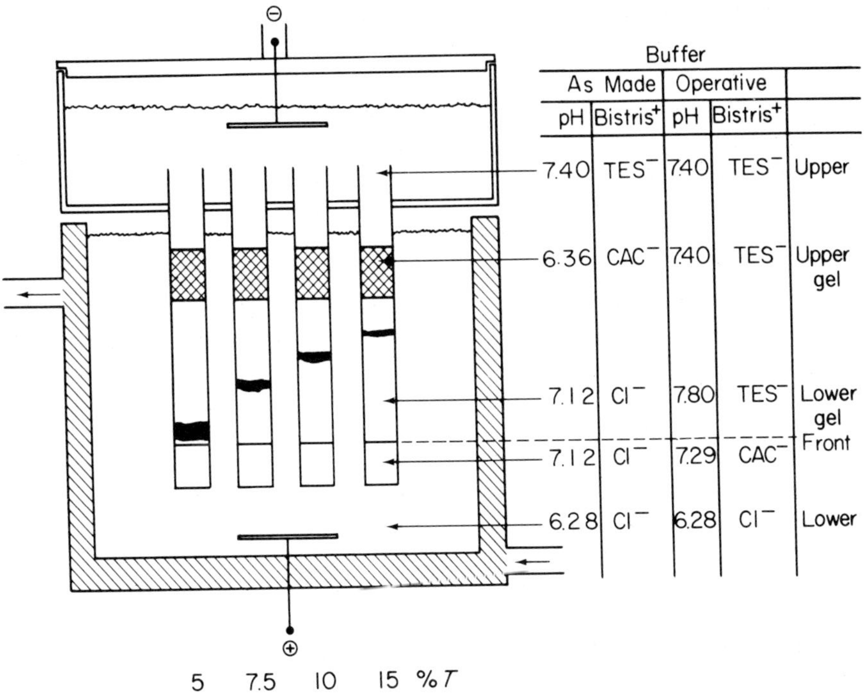

Fig. 2-10. Schematic representation of apparatus for analytical polyacrylamide gel electrophoresis. Polyacrylamide gels of 5, 7.5, 10, and 15 percent are shown. Bovine serum albumin is fractionated. Characteristic pH values for the various initial and operative buffer phases are shown. Stacking gels are indicated by the cross-hatched areas above the separation gel photographs. Upper and lower buffers in their respective resevoirs are shown with waved surfaces. Coolant within the jacket of the lower resevoir is represented by stripes, and the direction of coolant flow in indicated by the short arrows. BISTRIS is bis-(2-hydroxyethyl) iminotris (hydroxymethyl) methane; TES is *N*-tris (hydroxymethyl)-methyl-2-aminomethane-sulfonic acid; CAC is cacodylic acid; Cl is hydrochloride acid; % *T* is gel concentration. [From A. Chrambach and D. Rodbard, *Science*, **172**, 440 (1971). Copyright © 1971 by the American Association for the Advancement of Science.]

treated with a fixing agent to immobilize the proteins, and the location of the protein bands is observed after straining the gel with a suitable protein dye.

The gel is formed on polymerization of acrylamide ($CH_2{=}CH{-}CO{-}NH_2$) and the cross-linking agent methylene bisacrylamide ($CH_2{=}CH{-}CO{-}NH{-}CH_2{-}NH{-}CO{-}CH{=}CH_2$) directly in glass tubes. Polymerization is initiated by such catalysts as ammonium persulfate or by photopolymerization with riboflavin. Polymerization is accelerated with the inclusion of tetramethylenediamine in the mixture. Pore size of gel particles is related to the degree of cross-linking, which, in turn, is dictated by the proportion of acrylamide and methylene bisacrylamide used for polymerization.

Separation of components by disc gel electrophoresis depends on both charge and size differences. The resolution of components is improved if the separating gel is prepared with a pore size appropriate for a given molecular weight range. For example, the molecular sieving effect works best if a gel of a relatively large pore size (7 to 7.5 percent cross-linked gels) is used for high molecular weight proteins (10,000 to 1,000,000) and a small pore size (15 to 30 percent cross-linked gels) for low molecular weight components (less than 10,000).

The sample is applied to the gel in a solution containing high concentrations of glycerol or sucrose; the high density of the solution makes it easier to obtain a sharp starting zone of the protein components. A feature unique to disc gel electrophoresis is the possibility of concentrating the sample into thin zones or discs in the stacking gel (upper gel) at the beginning of the electrophoresis. Cross-linking of the stacking gel is lower than that of the separating gel, and the pH and composition of the buffers also differ in the two types of gels (Fig. 2-10).

During electrophoresis, a discontinuous voltage gradient is generated because of the ionic strength and pH differences. A boundary of fast-moving ions creates a "front," and a trailing zone of slow-moving ions follows behind. Protein ions migrate between these zones and are concentrated by the ion movement into thin zones. The concentrating effect operates only in the upper gel; and once the protein zones enter the lower gel, separation of the components begins, based on differences in charge and size. The excellent resolution of the system is due to the production of thin zones of sample in the stacking gel.

When the sample is added to the upper layer, a tracking dye of bromophenol blue is also included. The tracking dye has a mobility that is independent of pH; consequently, it moves with the anion front and marks the progress of electrophoresis. The position of the tracking dye forms a convenient reference boundary. The mobility of the components is related to the "front," which is indicated by the position of the tracking dye (Fig. 2-10).

Although disc gel electrophoresis serves admirably as an excellent test of purity, the method is quite versatile and is widely used in other applications. The method can be scaled upward and developed as an efficient procedure for preparative purposes, at least for the purification of samples in the range of up to about 100 mg. The analytical method serves to enumerate the number of protein components in an isolated cell structure and is useful in the characterization of isoenzymes—two or more genetically determined enzymes from a single organism with enzymatic properties in common but differing in their protein structures. It is also a direct test of whether two similar proteins are identical, at least in terms of the net charge and size of the molecules. From the point of view of following a purification procedure, disc gel electrophoresis serves to monitor the amount of contaminating protein components remaining and of indicating altered mobility due to chemical modification.

REFERENCES

Books

W. B. Jakoby (Ed.), "Enzyme Purification and Related Techniques," *Methods in Enzymology*, Vol. 22. Academic Press, New York, 1971.

M. Dixon and E. C. Webb, *Enzymes*, 3rd ed. Academic Press, New York, 1967.

J. B. Nielands and P. K. Stumpf, *Outlines of Enzyme Chemistry*, 2nd ed., Chapters 4 and 5. Wiley, New York, 1958.

Reviews and Articles

P. Cuatrecasas, "Affinity Chromatography of Macromolecules," *Advances Enzymol.* **36**, 29 (1972).

A. Chrambach and D. Rodbard, "Polyacrylamide Gel Electrophoresis," *Science*, **172**, 440 (1971).

H. A. Sober, R. W. Hartley, Jr., W. R. Carroll, and E. A. Peterson, "Fractionation of Proteins," *The Proteins*, **3**, 2 (1965).

J. Porath, "Cross-linked Dextrans as Molecular Sieves," *Advances Protein Chem.*, **17**, 209 (1962).

M. Dixon and E. C. Webb, "Enzyme Fractionation by Salting-Out: A Theoretical Note," *Advances Protein Chem.*, **16**, 197 (1961).

J. F. Taylor, "The Isolation of Proteins," *The Proteins*, **1**, 1 (1953).

Molecular Weight Determinations

THREE

The macromolecular nature of proteins, particularly their large size and instability toward certain experimental conditions, eliminates the use of methods that are usually employed for the determination of molecular weights of small molecules. Instead, new techniques that are suitable for use with these large molecules have been developed.

CHEMICAL ANALYSIS: MINIMUM MOLECULAR WEIGHTS

Chemical analysis provides data suitable for the calculation of a minimum molecular weight of a protein but not its true molecular weight. Pertinent analytical data useful for these calculations may be the content of individual amino acids (Chap. 4), the sulfur and phosphorus content, and the amount of a metal or prosthetic group tightly bound to the protein molecule. The minimum molecular weight is calculated on the assumption that one molecule of protein contains no less than one formula unit of the specified constituent. The calculation is based on the relationship

$$\text{Minimum M. W.} = \frac{\text{formula weight of constituent}}{\text{percent of constituent found}} \times 100 \qquad (3\text{-}1)$$

For example, the heme protein cytochrome c contains 0.43 percent iron (0.43 gram of iron per 100 grams of protein). With the use of Eq. (3-1), a value of

13,000 is calculated for the minimum molecular weight of cytochrome c. Since cytochrome c is known to contain only one iron atom per protein molecule, the minimum molecular weight in this case is equivalent to the true molecular weight.

The amount of an amino acid present in a protein may be used to calculate a value for the minimum molecular weight, a calculation that is performed for each of the amino acids. For example, bovine ribonuclease is found to contain 4.2 percent histidine from quantitative amino acid analysis of hydrolyzed samples (Chap. 4). The minimum molecular weight of the protein based on this value is

$$\frac{155 \times 100}{4.2} = 3690$$

Similar calculations for each one of the amino acids present in ribonuclease give values for the minimum molecular weight ranging from 7000 to 900. The reason for the range of values is related to the content of each amino acid, some occurring many times in the sequence and others less so, or only rarely. The relationship between the number of amino acid residues (amino acid content) and the molecular weight is shown in Eq. (3-2).

$$\text{True M. W.} = (\text{minimum M. W.}) \times (\text{number of residues per molecule}) \quad (3\text{-}2)$$

From a number of physical-chemical studies on the bovine ribonuclease molecule, the best estimate of the true molecular weight is 13,680. The number of residues of histidine is 13,680/3690, or 3.7. The nearest whole number, 4, is taken as the correct number of histidine residues present in the protein molecule. Similar calculations with each of the minimum molecular weights will give the number of residues of the corresponding amino acid. Since the number of residues of most amino acids is typically high, the minimum molecular weights are seldom used to estimate the true molecular weight. Instead the true molecular weight and the amino acid composition of a protein are used to calculate the number of amino acid residues present in the protein (see Chap. 4).

The true molecular weight of a protein must be obtained from a study of a physical property of the molecule, a property dependent on the size and shape of the macromolecule. For example, the behavior of a protein in the ultracentrifuge or on gel filtration is influenced by the size of the molecule; accordingly, both methods are useful for molecular weight estimations. Indeed, we will restrict our discussion primarily to these techniques, for they are widely used and they illustrate the usual approach taken by most investigators. In passing, it should be noted, however, that the molecular weight can also be obtained from the measurement of a colligative property, a value that depends directly on the number of molecules in solution. Two examples of this type are osmotic pressure and light scattering.

ULTRACENTRIFUGATION

Of the various methods available for molecular weight determinations, the use of the ultracentrifuge has been the method of choice for many years. An ultracentrifuge was first constructed by Svedberg in 1925 and used by him to demonstrate the suitability of the instrument for the study of macromolecules. In the intervening years, improvements in the methodology have brought the instrument and the technique to their present high level of sophistication.

The heart of the ultracentrifuge is a rotor designed to operate at speeds of up to 70,000 rpm, equivalent to a centrifugal force as high as 500,000 times gravity. The rotor accommodates a sector-shaped sample cell with quartz or sapphire windows so that optical measurements may be made during the sedimentation process (Fig. 3-1). Generally the optical system measures the refractive index of the solution, which is directly related to the concentration of the protein. Higher

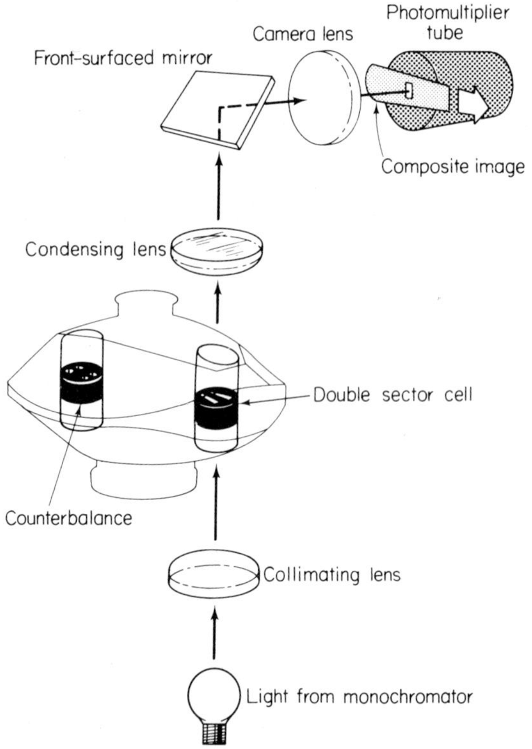

Fig. 3-1. A cross-section of the rotor and cell compartment of the ultracentrifuge. The rotor is mounted on a flexible shaft and rotated by means of a high-speed motor. The sample cell is positioned in the light path of the optical system and measurements are made during the centrifugation. [Illustration courtesy of Beckman Instruments, Inc.]

sensitivity, to detect low-protein concentrations, is obtained with an optical system incorporating ultraviolet absorption measurements, which is used at the absorption maxima (280 nm) of the aromatic residues. The temperature of the sample cell is held constant, and the rotor is contained in an evacuated chamber to minimize frictional resistance.

Sedimentation-velocity

In the sedimentation-velocity method, the ultracentrifuge is operated at high angular velocities so that the protein is subjected to a strong centrifugal force that drives the particles to the bottom of the cell. The centrifugal force is opposed both by the frictional resistance generated by the protein molecules as they move past solvent molecules and by the buoyancy experienced by them. The protein molecules sediment through the medium at a constant acceleration governed by the forces just described. The sedimentation process is normally followed by observing the rate of movement of the protein boundary. The boundary is the zone between solvent molecules at the meniscus (or close to the top of the cell) and the solution of protein molecules (Fig. 3-2). The movement of the boundary is followed with one of the several optical systems available.

The data obtained from sedimentation-velocity studies represent the rate of movement of the protein boundary, dx/dt, at a given centrifugal force, $\omega^2 x$. The value of x is the distance from the center of rotation to the midpoint of the protein boundary (in centimeters), t is the time (in seconds), and ω is the angular velocity in radians per second (2π times the number of revolutions of the rotor per second). The sedimentation coefficient, defined as the ratio of the velocity of the protein boundary to the centrifugal force, is given by the following expression:

$$s = \frac{dx/dt}{\omega^2 x} \tag{3-3}$$

The sedimentation coefficient is determined at several protein concentrations, and these values are extrapolated to give a corrected coefficient at zero-protein concentration. The "ideal behavior" expected at zero concentration minimizes protein interactions common at higher concentrations and gives a value of s typical of the monomer species.

The molecular weight of the protein is calculated from the Svedberg equation

$$M = \frac{RTs}{D(1 - \bar{v}\rho)} \tag{3-4}$$

where R is the gas constant (8.314×10^7 ergs per mole per degree), T is the absolute temperature in degrees Kelvin, s is the extrapolated sedimentation co-

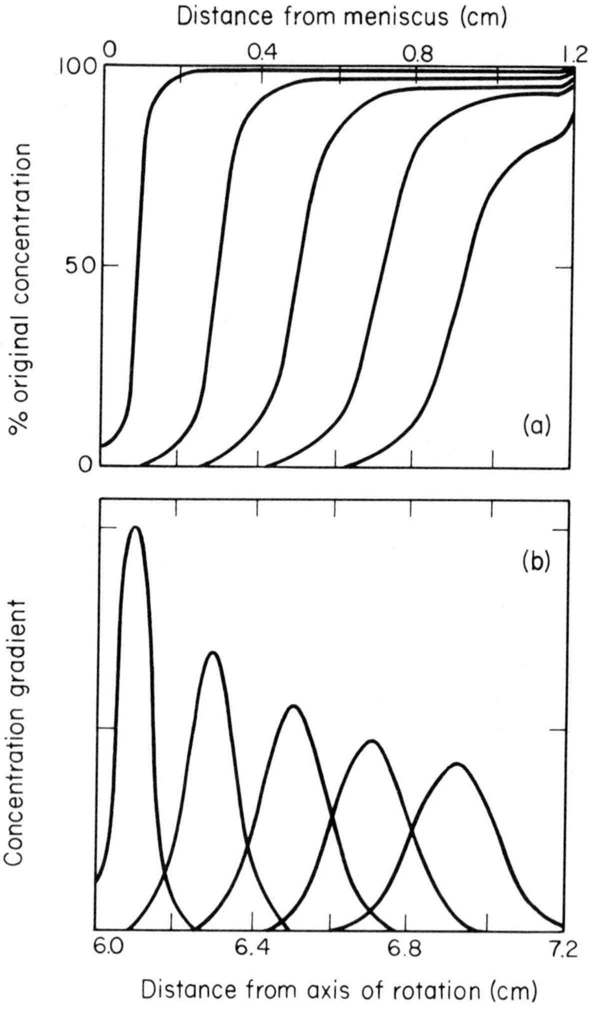

Fig. 3-2. A plot of protein concentration (c) and the concentration gradient (dc/dx) versus distance (x) from the axis of rotation during a sedimentation-velocity experiment. [From H. K. Schachman, *Methods in Enzymology*, 4, 32 (1957).]

efficient at zero-protein concentration,[1] D is the diffusion constant (centimeters squared per second), $\bar{v}$ is the partial specific volume of the protein molecule (cubic centimeters per gram), and ρ is the density of the solvent (grams

[1]The usual values of s are close to 1×10^{-13} sec. Therefore the designation of a Svedberg unit (S) is common practice where one unit is defined as 1×10^{-13} sec. Thus the sedimentation coefficient of lysozyme is 1.91×10^{-13} sec (Table 3-1), which is equivalent to 1.91 S.

per cubic centimeter). The experimentally derived parameters are corrected for temperature effects and expressed as the value they would have at 20°C.

The molecular weight of the protein molecule is directly proportional to the sedimentation coefficient, a parameter that is determined from measurements in the ultracentrifuge and is inversely related to the diffusion coefficient (Table 3-1). The diffusion coefficient is a measure of the transport process of protein particles in an undisturbed system, and its value is related to the size and shape of the protein molecule.

The diffusion coefficient is related to the flow of a component as given by Fick's equation (3-5) for the first law of diffusion.

$$\frac{dS}{dt} = -D \cdot A\frac{dc}{dx} \tag{3-5}$$

Here dS/dt is the amount of the component diffusing per second (in moles or grams per cubic centimeter) perpendicular to the cross-sectional area, A (in centimeters squared), of the diffusion cell, and dc/dx is the concentration gradient. Experimentally, one determines the diffusion coefficient in a system capable of following the movement of a protein boundary from a region of high concentration to a region of zero concentration.

The partial specific volume is the volume of solvent displaced by 1 gram of protein when added to a large volume of water. The value of $\bar{v}$ is determined experimentally from density measurements in a pycnometer or is estimated by

TABLE 3-1

Molecular Weights, Sedimentation Coefficients, Diffusion Constants, and Partial Specific Volumes for Several Proteins[a]

Protein	*Molecular Weight*	$s_{20,w} \times 10^{13}$	$D^0_{20,w} \times 10^7$	$\bar{v}$
Ribonuclease	13,700	2.00	11.9	0.728
Lysozyme	14,300	1.91	10.4	0.688
Chymotrypsinogen	25,800	2.54	9.5	0.721
β-Lactoglobulin	35,000	2.83	7.82	0.751
Ovalbumin	45,000	3.55	7.76	0.748
Serum albumin	65,000	4.60	5.94	0.734
Hemoglobin	68,000	4.13	6.9	0.749
Catalase	250,000	11.30	4.1	0.73
Urease	480,000	18.60	3.46	0.73
Tropomyosin	93,000	–	2.24	0.71
Fibrinogen	330,000	7.63	2.02	0.710
Collagen	345,000	3.05	0.69	0.695
Myosin	493,000	6.43	1.16	0.728

[a]Adapted from C. Tanford, *Physical Chemistry of Macromolecules*, Wiley, New York, 1961; and H. A. Sober (Ed.), *Handbook of Biochemistry*, 2nd ed., The Chemical Rubber Co., Cleveland, 1970.

calculations using the known $\bar{v}$ for each amino acid in conjunction with the amino acid composition of the protein. Most values of $\bar{v}$ fall within the range of 0.70 to 0.75 cm^3/g for globular proteins. The density of the solvent, ρ, must be obtained from experimental data and corrected to a temperature of 20 degrees.

In the Svedberg equation (3-4), the expression in parenthesis, $1 - \bar{v}\rho$, is a correction factor for the buoyancy of the protein particles in the solvent. With the usual value of $\bar{v}$ given above and with ρ close to one, the buoyancy factor becomes a difference between two numbers both close to one. As a result, a small error in $\bar{v}$ creates a much larger error in the calculated molecular weight.

Sedimentation-equilibrium

A somewhat different method for the determination of molecular weights makes use of the ultracentrifuge at moderate centrifugal forces, an analysis referred to as the sedimentation-equilibrium method. At low centrifugal forces, the movement of molecules toward the bottom of the cell is counterbalanced by diffusion to regions of low concentration—that is, toward the top of the cell. Measurements are made when the two opposing processes reach an equilibrium state, a condition characterized by an invariant concentration gradient (dc/dx) through the cell (Fig. 3-3).

The preceding conditions and assumptions form the basis for the derivation of a thermodynamic relationship describing the concentration gradient at different positions in the cell when equilibrium is reached.

$$\frac{1}{c}\frac{dc}{dx} = \frac{\omega^2 x M(1 - \bar{v}\rho)}{RT} \tag{3-6}$$

It is necessary to measure the protein concentration c at position x with an optical system capable of measuring refractive index changes or ultraviolet absorption. Both of these optical measurements are directly related to the protein concentration. The remaining symbols appearing in Eq. (3-6) are the same as those defined for Eq. (3-4).

The integrated form of the sedimentation-equilibrium equation is given in Eq. (3-7).

$$\ln \frac{c_x}{c_a} = \frac{\omega^2 M(1 - \bar{v}\rho)(x^2 - a^2)}{2RT} \tag{3-7}$$

The concentration of protein is shown as c_x and c_a, the values obtained at any position in the cell (x) and at the meniscus (a), respectively. When a single component is present, a plot of $\ln c$ as a function of x^2 should be linear (Fig. 3-4); the value of the slope is used in the calculation of the molecular

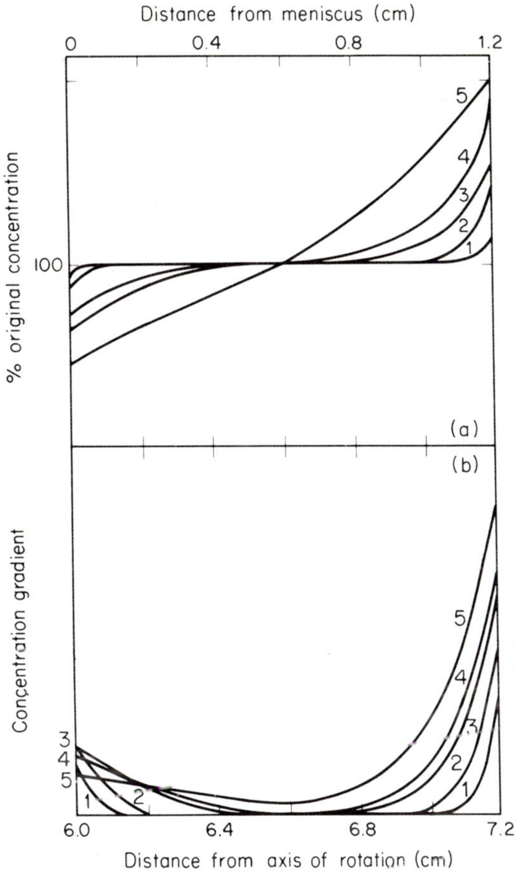

Fig. 3-3. A plot of protein concentration (c) and the concentration gradient (dc/dx) versus distance (x) from the axis of rotation during a sedimentation-equilibrium experiment. The numbers on the curves range from 1 to 5 and correspond to increasing time. Plot 5 shows the equilibrium distribution. [From H. K. Schachman, *Methods in Enzymology*, **4**, 32 (1957).]

weight [the slope is equated to the expression $\omega^2 M(1 - \bar{v}\rho)/2RT$, and the only unknown value is M].

A major problem associated with sedimentation-equilibrium experiments is the difficulty of determining the protein concentration when the system reaches equilibrium. Normally the refractive index system is used, but this can only give the difference in concentration between a specific point in the cell and some other point, such as the meniscus. Two approaches were developed to overcome this problem.

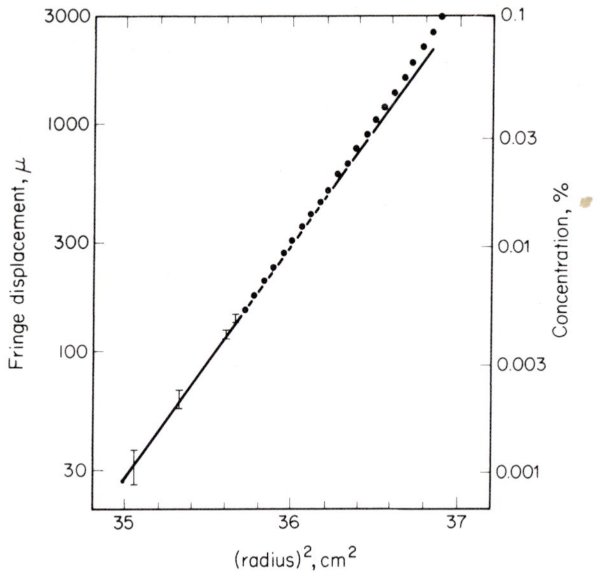

Fig. 3-4. A sedimentation-equilibrium experiment with bovine plasma albumin. The data are from a run of 0.05% protein at pH 4.4, 24.7 degrees for 14 hours at 24,630 rpm. The ordinate is the protein concentration on a logarithmic scale of the Rayleigh interference fringes and the corresponding concentration. [Reprinted from D. A. Yphantis, *Biochemistry*, 3, 297 (1964). Copyright © 1964 by the American Chemical Society. Reprinted by permission of the copyright owner.]

In the so-called low-speed method, a separate experiment is performed with a solution of the protein at a known concentration that serves to "calibrate" the refractive index changes. This step is accomplished by forming a synthetic boundary in the centrifuge cell, which produces a sharp protein boundary. In this way the refractive index changes may be observed as Rayleigh interference fringes; the number or displacement of the fringes with respect to their position in pure solvent is a measure of the initial protein concentration (c_0). After the calibration is completed, the actual sedimentation-equilibrium experiment is performed. The concentration of the protein is now determined, as the concentration difference (c_a and c_b), at the meniscus (at a) and the bottom of the solution column (at b). A slight variation of the equilibrium equation is used for the treatment of the data.

$$\frac{c_b - c_a}{c_0} = \frac{\omega^2 M(1 - \bar{v}\rho)(b^2 - a^2)}{2RT} \tag{3-8}$$

The "high-speed" method, also known as the "Yphantis method," was developed by Yphantis as a simple approach to determine the absolute concentration difference of the protein between any point in the cell and the meniscus. A high rotor speed is used to deplete the region near the meniscus of the protein (c_a becomes zero). Thus the difference in concentration of the protein at positions x and a gives the true concentration directly. These data are used in a plot of ln c as a function of x^2. In addition to the calculation of the molecular weight, the ln c vs x^2 plot serves as a test of the homogeneity of the sample. If a straight-line relationship is observed for all positions throughout the cell, the sample is considered to show a weight-average homogeneity.

Several days of centrifugation at a constant velocity may be required before the system reaches an equilibrium state. The time can be shortened to 24 hours or less with the use of a short column of solution (about 3 mm or less) in the direction of sedimentation.

A modification of the sedimentation–equilibrium procedure was proposed by Archibald as a means of completing the study in a matter of a few hours. In this approach to equilibrium procedure, the concentration of protein at different positions within the cell is determined in the early period of centrifugation before equilibrium is reached. Although a state of equilibrium cannot be reached in a short time period throughout the cell, the region at the meniscus and at the bottom of the cell is at equilibrium, for the sedimentation of particles in these regions is balanced by the diffusion process. Thus a restricted region of the cell can be used for the estimation of the protein concentration, and these data are used in the calculation of the molecular weight.

The sedimentation–equilibrium procedure is probably the most accurate of the methods presently available for the direct determination of the molecular weight of proteins. The method is suitable with molecules differing in sizes that range from a few hundred to several million, and it is applicable with molecules of varying shape, whether spherical, ellipsoidal, or rodlike (Table 3-2). The

TABLE 3-2

Molecular Weights Obtained from Sedimentation-Equilibrium Compared with Those from the Chemical Formula[a]

Protein	*From Chemical Formula*	*From Sedimentation-Equilibrium*
Sucrose	342.3	341.5
Ribonuclease	13,683	13,740
Lysozyme	14,305	14,500
Chymotrypsinogen	25,767	25,670

[a]From K. E. Van Holde, *Physical Biochemistry*, Prentice-Hall, Englewood Cliffs, N. J., 1971, p. 113.

sedimentation-velocity method is also versatile and highly reliable, but molecular weight values are less precise because the calculations depend on measurements of sedimentation as well as diffusion.

The study of protein molecules in the ultracentrifuge also gives useful information pertaining to their possible subunit structure. If a protein consisting of more than one polypeptide chain is submitted to either 8 *M* urea, 6 *M* guanidine hydrochloride, 1 percent sodium dodecyl sulfate, a high or low pH, or to low ionic strengths, the molecule may be dissociated to smaller-size particles if the polypeptide chains are held together solely by noncovalent interactions. If such is the case, the size of the subunits of the individual polypeptide chains may be determined from their altered sedimentation behavior. It is possible to recognize whether the subunits are all of one size or whether more than one type of subunit is present (Table 3-3).

TABLE 3-3

Subunit Constitution of Proteins[a]

Protein	*Molecular Weight*	*Subunits* *Number*	*Subunits* *Molecular Weight*
Insulin	11,466	2	5,733
β-Lactoglobulin	35,000	2	17,500
Chicken glycerol 3-phosphate dehydrogenase	60,000	2	30,000
Hemoglobin	64,500	4	16,000
Liver alcohol dehydrogenase	80,000	2	41,000
Pyruvate dehydrogenase	90,000	2	45,000
Yeast hexokinase	102,000	2	51,000
Glutathione reductase	124,000	2	56,000
Pig nucleoside diphosphokinase	138,000	6	21,000
Muscle aldolase	160,000	4	40,000
Glycerol kinase	217,000	4	55,000
Catalase	232,000	4	57,500
Glutamate dehydrogenase	320,000	6	57,000
Propionyl carboxylase	700,000	4	175,000

[a]Adapted from D. W. Darnell and I. M. Klotz, *Archives of Biochemistry and Biophysics*, **149**, 1–14 (1972).

MOLECULAR SIZE BY GEL FILTRATION

An approach that is completely different from that of the ultracentrifuge but quite suitable for the estimation of molecular weights is a chromatographic procedure known as gel filtration. An appropriate gel is selected as a solid support that is capable of acting as a molecular sieve, permitting some molecules to penetrate the sieve while excluding others (Fig. 3-5). The molecular sieve differentiates molecules based on their Stokes' radii, an empirical measure of the effective size of the protein molecule.

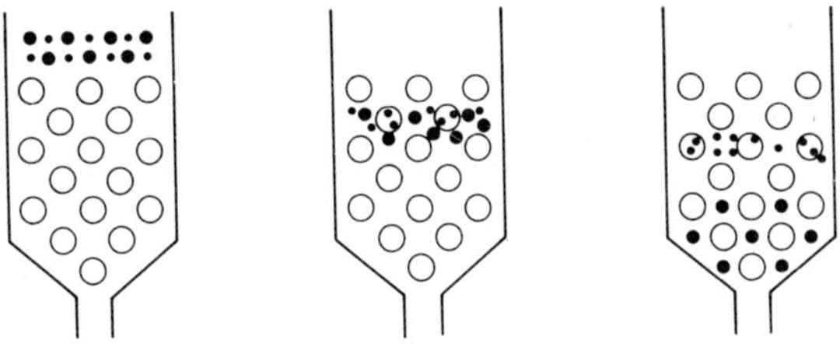

Fig. 3-5. The distribution of large molecules (large dots) and small molecules (small dots) at the start and at two later periods showing the separation of components on a column of a Sephadex gel (open circles).

Molecular sieves suitable for gel filtration are high molecular weight polymers of cross-linked dextrans, polyacrylamide resins, or agarose. The cross-linking creates a three-dimensional network with a pore size that limits the access of molecules into the interior. A given pore size allows molecules falling within a limited range of size and shape to penetrate the gel particles while all molecules of greater size are excluded (Fig. 3-6). The larger molecules that are excluded from the interior of the gels appear in the effluent fractions earlier than smaller molecules.

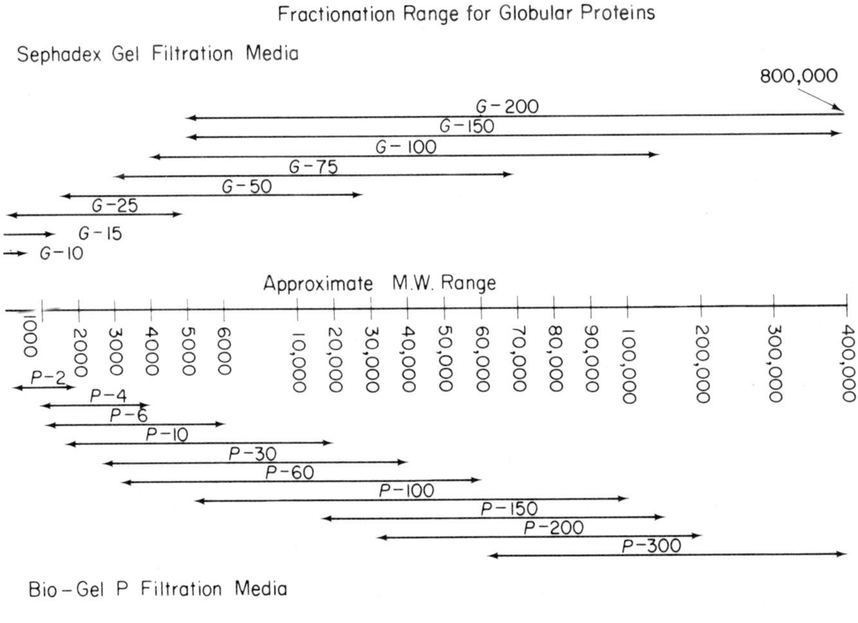

Fig. 3-6. Fractionation range for globular proteins on Sephadex gel filtration media and Bio-Gel P filtration media.

In order to perform a molecular weight determination on a sample, a gel filtration column must be calibrated with a set of proteins of known molecular weight. One plots the logarithm of the molecular weight of the protein and its relative elution volume, and a linear relationship is obtained (Fig. 3-7). The molecular sieve and the known proteins are chosen so that the unknown protein will fall within the calibrated range. With the selection of appropriate systems, molecular weights ranging from 2000 to 200,000 can be estimated with a precision of 10 percent.

Gel filtration as a means of determining the molecular size is fast, simple, and economical, but it lacks the high accuracy of ultracentrifugation studies. Nevertheless, its many desirable attributes are rapidly making gel filtration techniques a valuable adjunct to the more classical methods of molecular weight determinations. Furthermore, the technique works well for the estimation of the molecular weight of a single component in a mixture if a specific test for that component is available (for example, an enzyme assay or bioassay). Gel filtration is also useful for subunit analyses, for gel columns can be operated with solvents containing denaturing agents.

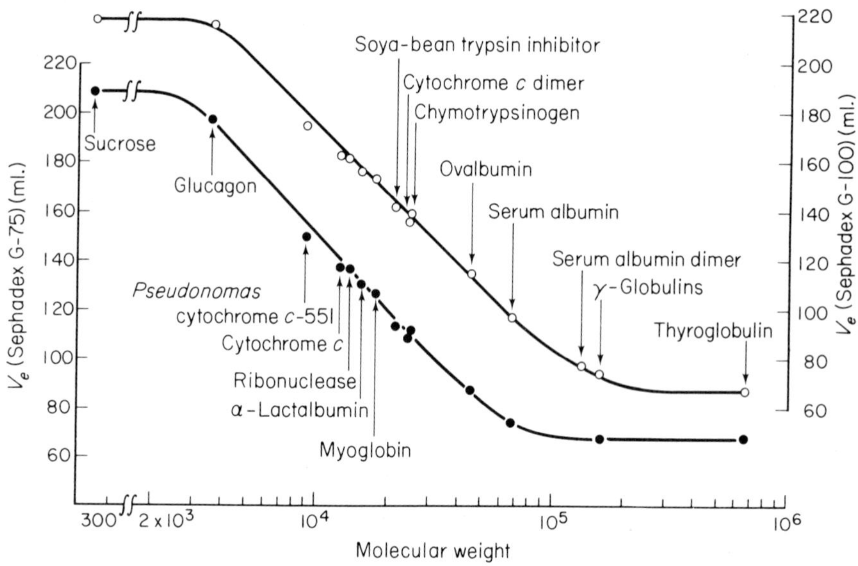

Fig. 3-7. Plots of elution volume, Ve, against the log molecular weight for a series of proteins on columns of Sephadex G-75 (•) and G-100 (○). [From P. Andrews, *Biochem. J.*, **91**, 222 (1964).]

GEL ELECTROPHORESIS

Another recent development of a technique for molecular weight determinations at a microlevel is gel electrophoresis. An electrophoretic separation is carried out in small tubes (6 X 60 mm) containing a polyacrylamide gel of a given cross-linkage (Chap. 2). The protein is dissolved in a buffer at the desired pH, containing 0.1 percent of sodium dodecyl sulfate and 0.1 percent of β-mercaptoethanol. The buffer solution is also used to adjust the gel to these conditions. The sample is applied to the top of the gel along with Bromphenol blue (tracking dye), which serves as a visible reference marker during the separation. Electrophoresis is performed at a low current for approximately 4 hours, or for the time required for the tracking dye to move far down the tube. At the conclusion of the electrophoretic separation, the gel is removed from the tube and the position of the protein band is revealed by staining the protein with a dye (Fig. 3-8).

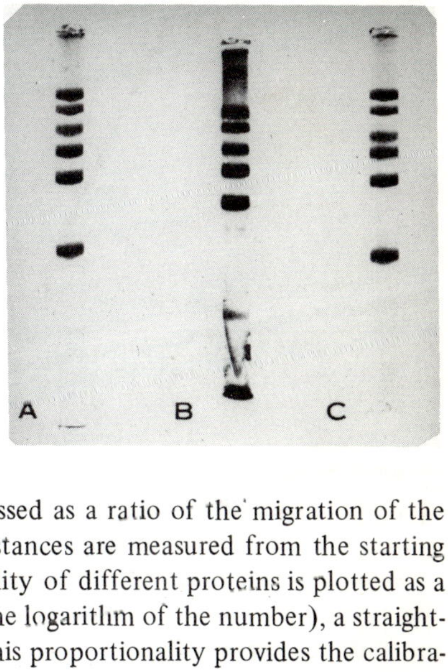

Fig. 3-8. Separation of the polypeptide chains of different proteins. The proteins are listed from top to bottom, (a), catalase (60,000), fumarase (49,000), aldolase (40,000), glyceraldehyde phosphate dehydrogenase (36,000), carbonic anhydrase (29,000), and myoglobin (17,200); (b), same as (a), but omitting the myoglobin. This gel was run on a different occasion; (c), catalase, fumarase, liver alcohol dehydrogenase (41,000), glyceraldehyde phosphate dehydrogenase, carbonic anhydrase, and myoglobin. [From K. Weber and M. Osborn, *J. Biol. Chem.*, **244**, 4406 (1969).]

The mobility of the protein is expressed as a ratio of the migration of the protein to that of the dye zone; both distances are measured from the starting position. When the electrophoretic mobility of different proteins is plotted as a function of their molecular weights (as the logarithm of the number), a straight-line relationship is observed (Fig. 3-9). This proportionality provides the calibration necessary to use gel electrophoresis as a method for molecular weight estimations.

Apparently the presence of the anionic detergent, sodium dodecyl sulfate, masks the charge differences that exist with different proteins, and the mobility of the protein is therefore dependent only on the size of the molecule. The method is capable of distinguishing between proteins that may differ by only 10 percent in their molecular weights in the range from 15,000 to 100,000.

The technique of gel electrophoresis is rapidly becoming widespread in its acceptance as a simple, inexpensive, rapid, and extremely sensitive method for molecular weight studies. It is also useful for the study of oligomeric proteins because the electrophoresis is performed with a detergent present, which causes the dissociation of the protein to the subunit level. If mercaptoethanol is also

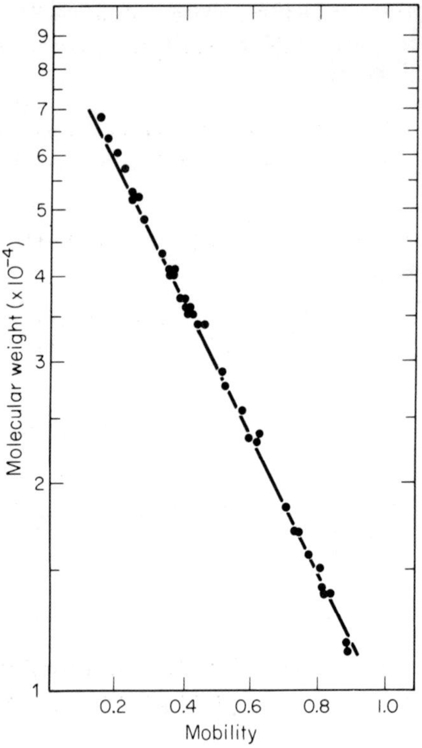

Fig. 3-9. The relationship of the molecular weight of a set of 37 different proteins to their mobility. The molecular weights range from 11,000 to 70,000. [From K. Weber and M. Osborn, *J. Biol. Chem.*, **244**, 4406 (1969).]

present, disulfide bonds are reduced and polypeptide chains separate if intermolecular disulfide bonds existed between the chains. Indeed, the behavior of a protein on gel electrophoresis after treatment with sodium dodecyl sulfate or sodium dodecyl sulfate and mercaptoethanol is a most useful approach in judging the extent of subunit structure (see Table 3-3 for a summary of examples of proteins with subunits).

REFERENCES

Books

T. J. Bowen and A. J. Rowe, *An Introduction to Ultracentrifugation.* Wiley-Interscience, New York, 1970.

J. H. Coates, "Ultracentrifugal Analysis," in *Physical Principles and Techniques of Protein Chemistry*, S. J. Leach (Ed.). Academic Press, New York, 1970, Part B.

K. E. Van Holde, "Physical Characterization of the Protein Molecule," in *Protein Sequence Determination*, S. B. Needleman (Ed.). Springer-Verlag, New York, 1970.

D. J. Winzor, "Analytical Gel Filtration," in *Physical Principles and Techniques of Protein Chemistry*, S. J. Leach (Ed.). Academic Press, New York, 1969, Part A.

J. W. Williams (Ed.), *Ultracentrifugal Analyses in Theory and Experiment.* Academic Press, New York, 1963.

S. Claesson and I. Moring-Claesson, "Ultracentrifugation," in *A Laboratory Manual of Analytical Methods of Protein Chemistry,* Vol. 3. Pergamon Press, New York, 1961.

H. K. Schachman, *Ultracentrifugation in Biochemistry.* Academic Press, New York, 1959.

T. Svedberg and K. O. Pederson, *The Ultracentrifuge.* Oxford University Press, New York, 1940.

Reviews and Articles

K. G. Mann and W. W. Fish, "Protein Polypeptide Chain Molecular Weights by Gel Chromatography in Guanidinium Chloride," *Methods in Enzymology*, **26**, 28 (1972).

P. Andrews, "Estimation of Molecular Size and Molecular Weights of Biological Compounds by Gel Filtration," *Methods of Biochem. Analysis*, **18**, 1 (1970).

G. K. Ackers, "Analytical Gel Chromatography of Proteins," *Advances Protein Chem.*, **24**, 343 (1970).

K. Weber and M. Osborn, "The Reliability of Molecular Weight Determinations by Dodecyl Sulfate-Polyacrylamide Gel Electrophoresis," *J. Biol. Chem.*, **244**, 4406 (1969).

P. Andrews, "Estimation of the Molecular Weight of Proteins by Sephadex Gil-Filtration," *Biochem J.*, **91**, 222 (1964) and **96**, 595 (1965).

H. K. Schachman, "The Ultracentrifuge: Problems and Prospects," *Biochem.*, **2**, 887 (1963).

J. W. Williams, K. E. Van Holde, R. L. Baldwin, and H. Fujita, "The Theory of Sedimentation Analysis," *Chem. Reviews*, **58**, 715 (1958).

Amino Acid Analysis

FOUR

The determination of the amino acid composition of a protein is the first step taken in characterizing the chemical structure of the molecule. Amino acid analysis directly provides two essential types of information: it completely describes a protein in terms of the amino acids present and it gives a precise value for the amount of each amino acid per unit mass of protein. However, the composition of a protein is best described as the number of residues of each amino acid per protein molecule. Conversion of the amino acid composition to residue numbers requires a knowledge of the molecular weight of the protein (Chap. 3).

OUTLINE OF METHODS

Current methods of amino acid analysis operate at the microgram level with a precision of 2 to 5 percent and require only a few hours for completion. The description of a complete amino acid analysis may be conveniently divided into a sequence of consecutive operations. First, the protein is hydrolyzed to a mixture of amino acids. Second, the amino acid mixture is separated into individual components by a method utilizing a system of high resolving power. Lastly, the quantity of each amino acid is determined by colorimetric analysis or with other detection systems and the data are used to calculate the molar amount of each amino acid. These experimental operations are examined in detail.

HYDROLYSIS OF PROTEINS

Acid hydrolysis

The hydrolysis of a protein is catalyzed by mineral acids at an elevated temperature. Constant boiling hydrochloric acid or 6 *N* HCl is the acid of choice because of its high purity, low price, and ease of removal at the end of the hydrolytic step. A controlled temperature of 105 or 110 degrees is used, and samples are hydrolyzed for periods of 20 to 96 hours, or possibly longer. The sample is hydrolyzed in the absence of air in order to eliminate the destruction of certain amino acids.

The accuracy of amino acid analysis is influenced by two problems associated with the hydrolytic process. The first is the slow rate of hydrolysis of peptide bonds of amino acid residues with bulky side chains. For example, the rates of hydrolysis of valyl, isoleucyl, and tryptophanyl residues are slower than those of glycyl residues. Bulky side chains shield their peptide bonds, and the steric hindrance decreases the concentration of hydronium ions in the vicinity of the peptide bond. In early periods of hydrolysis (20 hours), the yield of these amino acids is low, and hydrolysis times of 70 to 90 hours are required for their complete release.

The second problem associated with acid hydrolysis is the acid lability of a few of the amino acids under the usual conditions used for the hydrolysis. Serine, threonine, cysteine, and a few others are converted in part to other compounds, and an incomplete recovery is found after 20 hours of hydrolysis. Tryptophan is so extensively destroyed on acid hydrolysis that other methods must be used for its analysis (see below).

Most of the limitations imposed by acid hydrolysis can be overcome by determining the amino acid content after several hydrolysis times. Usually duplicate or triplicate analyses are performed after 24, 48, 72, and 96 hours of hydrolysis. These data provide multiple values for the amino acid content, and an average value is calculated for those amino acids that remain constant with time. The amino acid residues showing slow release are estimated from their values after 70 or 96 hours. A typical progress curve is shown in Figure 4-1, and it is clear that valine and isoleucine are best evaluated after an extended hydrolysis period where the maximal value for their recovery is used. The content of acid-labile amino acids, such as serine and threonine, is estimated from extrapolation to their zero time values.

A method suitable for the determination of tryptophan utilizes *p*-toluenesulfonic acid as the catalyst for hydrolysis rather than mineral acid. The hydrolysis is carried out in 3 *N* *p*-toluenesulfonic acid containing 0.2 percent 3-(2-aminoethyl)-indole at 110 degrees in evacuated, sealed tubes. Addition of the indole derivative in the hydrolysis mixture gives a higher and more reproducible recovery of tryptophan. Generally it is again necessary to use hydrolysis times

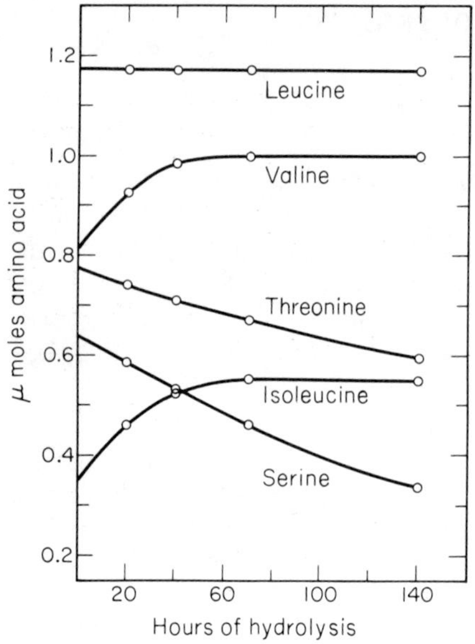

Fig. 4-1. Amino acid recoveries as a function of hydrolysis time. The ordinate gives the micromoles of amino acid recovered from 1.41 mg of rabbit muscle ATP-AMP transphosphorylase (myokinase). [From T. A. Mahowald, E. A. Noltmann, and S. A. Kuby, *J. Biol. Chem.*, **237**, 1138 (1961).]

of 22, 48, and 72 hours and to extrapolate the tryptophan values to zero time, assuming first-order kinetics. Analyses of several proteins for tryptophan gave recoveries that were close to the theoretical values, and the recoveries of the remaining amino acids were comparable to those obtained after hydrolysis with 6 *N* HCl.

Base hydrolysis

Most of the amino acids present in proteins are almost completely destroyed during alkaline hydrolysis; consequently, a complete amino acid analysis is not possible. Hydrolysis of a protein with alkali is restricted to the preparation of a sample suitable for the analysis of tryptophan and those few amino acid derivatives that are unstable in acid solutions.

The protein is maintained at 110 degrees for 16 hours in 4.2 *N* sodium hydroxide in sealed, evacuated tubes that are lined with polypropylene. Starch is included in the hydrolysis mixture as an antioxidant, for it improves the recovery of tryptophan. Trial experiments showed that the recoveries of tryptophan from purified proteins of known content were essentially quantitative when ion-exchange chromatography was used for the estimation of the amino acid. With some proteins, however, the rate of release of tryptophan may be slow, and a trial analysis after 98 hours of hydrolysis should also be performed.

Enzymic hydrolysis

The availability of a large number of highly purified proteolytic enzymes of differing specificity offers the possibility of completely hydrolyzing a protein with enzymes as catalysts. Indeed, a procedure directed toward this goal was developed several years ago and applied to a few well-characterized proteins. However, the technique has yet to see widespread acceptance. On the other hand, the complete enzymic hydrolysis of peptides is widely used in sequence studies, and the discussion that follows applies equally well to most peptides.

The protein or large peptide to be analyzed is first denatured with heat so that it becomes susceptible to proteolysis. The substrate is hydrolyzed with one or more endopeptidases to produce a mixture of small peptides. Suitable enzymes with a broad specificity are papain, pepsin, and subtilisin. Alternatively, the protein is first degraded with a mixture of enzymes, an ideal combination being trypsin and chymotrypsin. Finally, the small peptides that are produced are hydrolyzed completely to amino acids with aminopeptidase M or with a mixture of leucine aminopeptidase and prolidase (see Chap. 9).

Complete enzymic hydrolysis offers certain advantages over acid hydrolysis. Amino acids normally modified or destroyed by acid hydrolysis are obtained directly. Thus tryptophan, asparagine, and glutamine are released intact on enzymic hydrolysis. Furthermore, amino acid derivatives that are present in some proteins—serine phosphate or tyrosine sulfate are examples—are also liberated by enzymes.

Although the advantages of enzymic hydrolysis are attractive, several difficulties are inherent to the procedure. Hydrolysis of the protein substrate to a mixture of peptides becomes more difficult as the size of the protein increases. As hydrolysis proceeds, products of the reaction may inhibit the enzyme; and if the proteolytic enzyme denatures its activity will decrease with time. Finally, a potential danger is the release of amino acids from self-digestion of the proteolytic enzymes themselves. Some of these problems are eliminated or minimized if hydrolysis is performed with low concentrations of enzymes that display high catalytic efficiency.

SEPARATION OF AMINO ACIDS

As many as 20 different amino acids may be found in a protein hydrolysate, and suitable procedures must be available to separate each component. The separation of 20 components is in itself extremely difficult, and the task is further complicated by the similar structures of many of the amino acids, such as leucine and isoleucine, serine and threonine, or tyrosine and phenylalanine.

Ion-exchange chromatography

Thus the complete separation of all 20 amino acids requires methods of extremely high resolving power. Spackman, Stein, and Moore developed methods capable of resolving amino acid mixtures by ion-exchange chromatography. The following discussion will be restricted to the procedure developed by these investigators for automatic amino acid analysis.

The separation of amino acids is accomplished by chromatography on ion-exchange resins of the sulfonic-acid type. These are high molecular weight, insoluble polymers prepared by copolymerization of styrene with divinylbenzene followed by sulfonation of the styrene-divinylbenzene beads. The ion-exchange resin is a three-dimensional structure of inert hydrocarbons with a large number of functional sulfonic-acid groups. Commercially produced resins are available with different degrees of cross-linking, of variable particle sizes, and as either spherical beads or as crushed particles.

The ion-exchange process is represented by the reaction given in Eq. (4-1).

$$\text{Resin–SO}_3^-(\text{Na}^+) + {}^+\text{NH}_3\underset{}{\overset{\text{R}}{\overset{|}{\text{C}}}}\text{HCOOH} \rightleftharpoons \text{Resin–SO}_3^-({}^+\text{NH}_3\overset{\text{R}}{\overset{|}{\text{C}}}\text{HCOOH}) + \text{Na}^+ \quad (4\text{-}1)$$

An amino acid mixture, adjusted to pH 2.2, is placed on a column of the resin previously equilibrated with a sodium citrate buffer at pH 3.25. At a low pH and ionic strength, the amino acids displace sodium ions and bind to the sulfonic-acid groups by electrostatic interactions (see Chap. 2).

The strength of binding of amino acids to the resin is a function of the pH and ionic strength of the buffer; the order of elution from the column is related to the affinity of each amino acid for the resin. With an increase in pH, the α-carboxyl group becomes ionized, and the net charges of the amino acids approach zero for neutral amino acids, are negative for the acidic amino acids, and are positive for the basic amino acids. The ion-exchange reaction shifts toward the left as the pH increases [Eq. (4-1)], and the acidic amino acids are eluted first, followed by the neutrals, and finally by the basics. An increase in ionic strength also shifts the position of equilibrium to the left, since the increased concentration of sodium ions competes for sites on the resin. Suitable conditions of buffer composition, pH, ionic strength, and temperature are selected whereby the bound amino acids are selectively displaced from the resin in a sequential order.

The separation of amino acids on columns of ion-exchange resins is dominated by the ion-exchange process, but the separation is also influenced by adsorption of some of the amino acids to the resin; amino acids with nonpolar substituents are adsorbed to the hydrocarbon matrix by hydrophobic interactions, in addition to their usual electrostatic interactions with the sulfonic-acid groups.

The binding of these amino acids to the resin is greater than if ion-exchange was the sole process, and their elution volumes are correspondingly larger. Apparently these effects contribute to the observed order of displacement of serine and threonine, glycine and alanine, and leucine and isoleucine, to cite a few of the more important cases.

The present state of the art of column chromatography on ion-exchange resins is due to the systematic investigations of Stein and Moore, Piez, Hamilton, and others, on the influence of pH, ionic strength, temperature, flow rate, and resin type on the resolution of amino acid mixtures. Several systems were developed for the complete analysis of a protein hydrolysate on a single ion-exchange column or on two columns. New modifications were introduced to increase sensitivity, to decrease the time required for an analysis, and to automate the entire procedure.

Amino acid analyzer

The amino acid analyzer is commercially available as a complete unit housing all the components necessary for the separation, quantitation, and data recording of the amino acids found in proteins and in physiological fluids (Fig. 4-2).

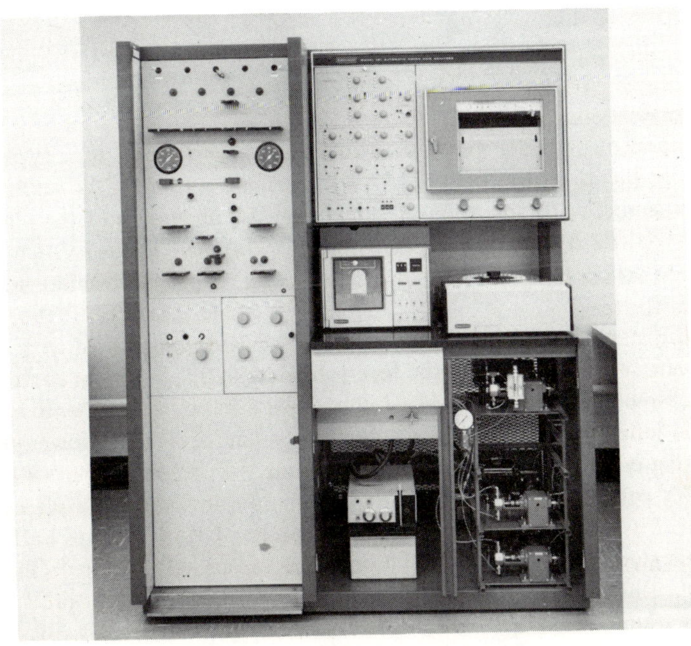

Fig. 4-2. A commercially available amino acid analyzer. The model 121, automatic analyzer. [Illustration courtesy of Beckman Instruments, Inc.]

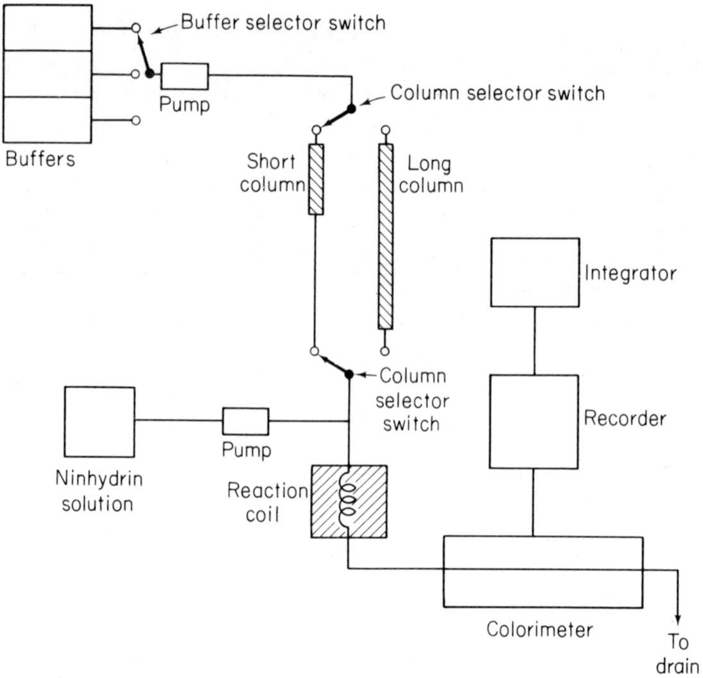

Fig. 4-3. The operation of the amino acid analyzer is shown as a flow diagram.

The component parts are shown as a flow diagram in Fig. 4-3 for a system that depends on the use of two columns. A short column is used for the resolution of the basic amino acids and a longer column for the separation of the remaining amino acids. Each column is operated with a different buffer system that is selected to achieve the maximum separation of the appropriate amino acid mixtures.

At the start of the analysis, the short column is equilibrated with 0.35 *N* sodium citrate at pH 5.28 and the long column with 0.2 *N* sodium citrate at pH 3.25. A sample of the amino acid mixture at pH 2.2 is introduced onto both the short and long columns. Since the solution of amino acids is at a lower pH than the starting buffers, the amino acids bind strongly to the resin [Eq. (4-1)]. As the buffer percolates through the column, amino acids become displaced from the resin. First, the short column is eluted with the 0.35 *N*, pH 5.28 buffer, and when this step is completed, the long-column separation is started (Fig. 4-4). The long column is operated initially with 0.2 *N* sodium citrate at pH 3.25, but the separation is completed with 0.2 *N* sodium citrate buffer at pH 4.28. A timing arrangement permits the automatic switching from the first buffer to the second. Buffer is introduced to the columns at a controlled flow rate by means of a high-precision liquid pump.

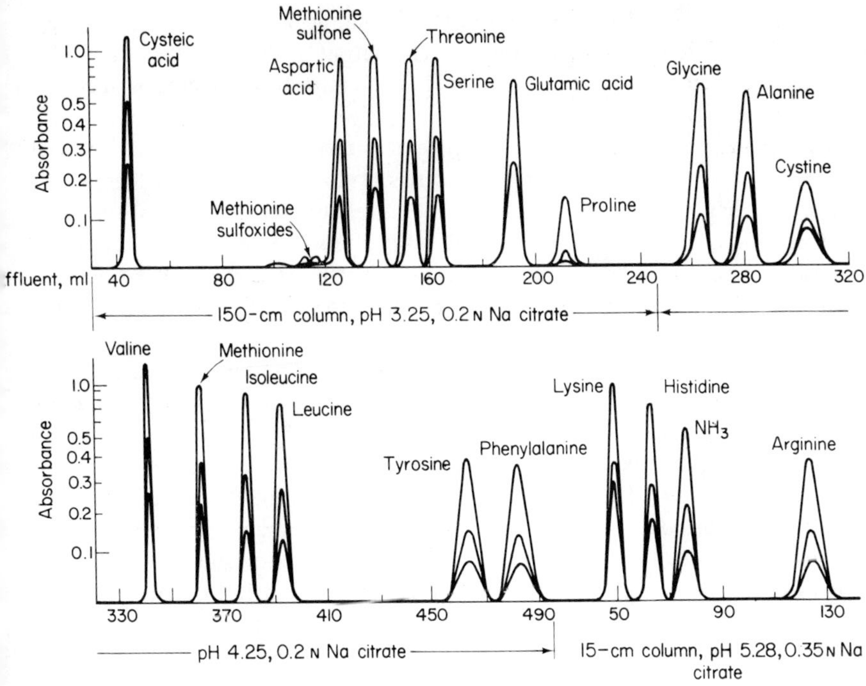

Fig. 4-4. Chromatographic separation of a synthetic mixture of amino acids on a sulfonated polystyrene resin. An automatic amino acid analyzer was used for the sample separation, the reaction with the ninhydrin reagent, and the automatic recording of the absorbance at 570 nm. [From D. H. Spackman, W. H. Stein, and S. Moore, *Anal. Chem.*, **30**, 1190 (1958). Copyright © 1958 by the American Chemical Society. Reprinted by permission of the copyright owner.]

The column effluent, as a flowing stream, enters a mixing manifold where it meets a second stream consisting of the ninhydrin reagent. The combined solution is pumped through a reaction coil that is immersed in a water bath maintained at 100°C. The time required for the solution to flow through the reaction coil is sufficient for the amino acids to react completely with ninhydrin so that a purple derivative is formed (described below).

The effluent stream emerging from the reaction coil is continuously monitored by a colorimeter placed in the path of the stream. The colorimeter consists of a flow-through cell made up of three cuvettes with individual light sources, filters, and photocells. The first cuvette is adjusted to a wavelength of 570 nm, which coincides with the absorption maximum of the purple color produced from the reaction of most of the amino acids with ninhydrin. The second cuvette is set to a wavelength of 440 nm, which is the absorption maximum of the yellow compound produced when proline reacts with ninhydrin. The response from the third cuvette is used when the size of an amino acid peak is large and exceeds the range of the first 570-nm reading. The output of the colorimeter provides a response for all amino acids encountered as constituents of proteins and covers a wide range of sample concentration.

The electrical output of the colorimeter is introduced to three independent channels of an electronic recorder. A series of dots are printed at precise intervals, and their position on the chart paper is related to the output of the photocells of the colorimeter. The record of an analysis is presented as a set of three overlapping curves with the absorbancy of the effluent stream plotted as a function of time and with each curve corresponding to the output of a single cuvette (Fig. 4-4).

Amino acid peaks are identified from calibration runs with known mixtures of amino acids. The reproducibility of elution volumes is excellent, for the flow rates for the buffer and ninhydrin solutions are accurately controlled. The amount of each amino acid is determined by measuring the area of the peak on the chromatogram. The symmetrical distribution of dots on either side of the peak maximum makes it possible to use a calculation based on the assumption that the area is approximately the same as the area of a triangle. The true height of the peak is multiplied by the width at half that height. Equation (4-2) ex-

$$\mu\text{moles of an amino acid} = \frac{H \times W}{C} \tag{4-2}$$

presses the concentration of the peak in terms of its area ($H \times W$) and a calibration constant (C). The calibration constant C is the peak area per μmole of the amino acid, a value experimentally determined from analyses of known amounts of the amino acid.

The amino acid analyzer, as described above, is a highly sophisticated instrument, and the basic unit is widely used throughout the world. A number of modifications that improve resolution, precision, sensitivity, and increase the level of automation have been introduced. New types of resins that are spherical beads of uniform particle size increase the resolution and operate well with higher flow rates. The time required for a complete analysis is decreased accordingly. A longer light path in the colorimeter and electronic amplification of the output increase the sensitivity of the ninhydrin analysis, permitting the use of a smaller sample. An automatic sample injector system and program-controlled operation of the analyzer transform the instrument to completely automatic operation and extend the useable time for analysis to a 24-hour basis. Current models of the analyzer are suitable for samples as small as 0.005 micromoles per amino acid, take 2 hours for a complete analysis, and ideally operate with a precision of ±2 percent.

A considerable decrease in the time required for calculations is gained with the use of an electronic integrator. This is an instrument that is attached to the amplifier of the recorder; it senses the change in slope of the colorimeter output and "integrates" the area of each peak. The data are presented as a digital readout on an attached printer.

NINHYDRIN REACTION

Reaction with α-amino acids

The development of a stable ninhydrin reagent by Moore and Stein led to its widespread acceptance for amino acid analyses and also for the detection and analysis of peptides and proteins. In the reaction of α-amino acids with ninhydrin, the amino acid is oxidatively deaminated to an aldehyde, carbon dioxide, and ammonia [Eq. (4-3)].

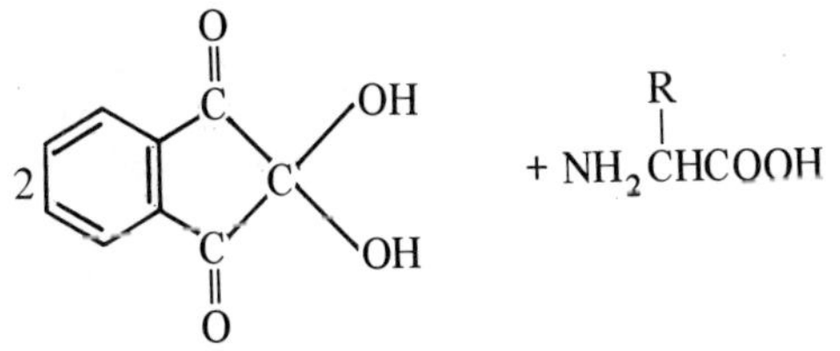

Ninhydrin (Triketohydrindene, or 2,2-dihydroxy-1,3-indandione)

α-Amino acids

(4-3)

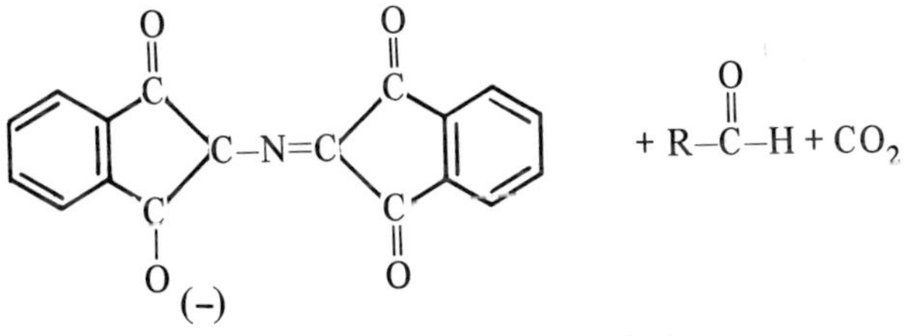

Ruhemann's purple

The ammonia and 2 moles of the reagent form a complex known as Ruhemann's purple or, in chemical terms, as diketodihydrinylidene-diketohydrindamine.

The maximum yield of the complex is attained with a reaction time of 15 minutes at 100 degrees and at pH 5. However, the yield is related to the stability of the reagent, and the solution must be protected from the atmosphere. The solution is protected by added stannous chloride, hydrindantin, which is the reduced form of ninhydrin, or by cyanide, and the solution is purged of dissolved air with nitrogen. The ninhydrin solution also contains a strong buffer sufficient to maintain the column effluent at pH 5.

The reaction product is stable for several hours, and the amount present is determined colorimetrically at the absorption maximum of 570 nm. Beer's law behavior is obeyed over a wide concentration range. Unfortunately, the color yield is not stoichiometric for all amino acids but is sufficiently high for most. The color yield of the amino acids must be measured (the calibration constant) and the appropriate value is used in all calculations required to complete an amino acid analysis.

Reaction with proline

The reaction of ninhydrin with imino acids, such as proline and hydroxyproline, leads to the formation of a different product [Eq. (4-4)].

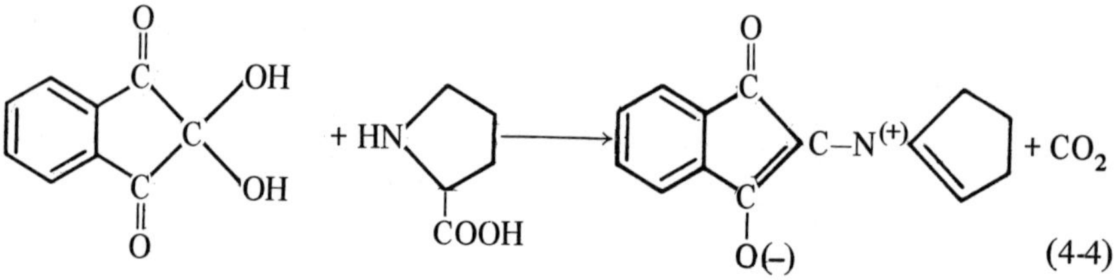

(4-4)

Ninhydrin　　Proline　　Yellow color

This compound is yellow rather than purple found with α-amino acids. The absorption maximum of the yellow product is 440 nm, so that a separate determination at this wavelength is required for the estimation of proline.

REFERENCES

Books

A. Niederwiesser and G. Pataki (Eds.), *New Techniques in Amino Acid, Peptide, and Protein Analysis.* Ann Arbor Science Publishers, Ann Arbor, 1971.

S. Blackburn, *Amino Acid Determination.* Marcel Dekker, Inc., New York, 1968.

C. J. O. R. Morris and P. Morris, *Separation Methods in Biochemistry.* Interscience Publishers, New York, 1963.

Reviews and Articles

T. E. Hugli and S. Moore, "Determination of the Tryptophan Content of Proteins by Ion-Exchange Chromatography of Alkaline Hydrolysates," *J. Biol. Chem.*, **247**, 2828 (1972).

T.-Y. Liu and Y. H. Chang, "Hydrolysis of Proteins with *p*-Toluenesulfonic Acid," *J. Biol. Chem.*, **246**, 2842 (1971).

A. Linenberg, "Gas Chromatography of Amino Acid Derivatives," in *Protein Sequence Determination*, S. B. Needleman (Ed.). Springer-Verlag, New York, 1970.

D. H. Spackman, "Accelerated Methods," *Methods in Enzymology*, **XI**, 3 (1967).

P. B. Hamilton, "Micro and Submicro Determinations of Amino Acids by Ion-Exchange Chromatography," *Methods in Enzymology*, **XI**, 15 (1967).

A. Light and E. L. Smith, "Amino Acid Analysis of Peptides and Proteins," *The Proteins*, **1**, 2 (1963).

G. R. Tristram and R. H. Smith, "The Amino Acid Composition of Some Purified Proteins," *Adv. Prot. Chem.*, **18**, 227 (1963).

D. H. Spackman, W. H. Stein, and S. Moore, "Automatic Recording Apparatus for Use in the Chromatography of Amino Acids," *Anal. Chem.*, **30**, 1190 (1958).

Sequence Studies Introductory Remarks

FIVE

At the present state of the art, the elucidation of the amino acid sequence of a typical protein containing 200 to 300 residues is still a Herculean task requiring several man-years of work. Nevertheless, the sequences of a large number of proteins have been completed, and one would predict that the rate of progress will accelerate in the near future. Within a decade or two of this writing, the sequence of most purified globular proteins will probably be known.

If one considers the nature of the polypeptide structure of a protein, the method of choice for sequence analysis would be a procedure that operates sequentially on the protein molecule. In this way the first residue, the second, third, and so forth, would be identified and the complete sequence would be established. Unfortunately, the complete sequence cannot be determined in this fashion at present, although rapid progress is being made with very encouraging results.

GENERAL CONSIDERATIONS

The strategy followed by most investigators is based on the one developed by Sanger in his pioneering research on the amino acid sequence of the polypeptide hormone insulin. Sequence studies use a multistep process that includes (a) the degradation of a protein to a mixture of overlapping peptides, (b) sequence analysis of each peptide, and (c) assembly of the individual peptides into a complete alignment of amino acid residues that correctly describe the primary structure of the protein molecule.

In the beginning studies on the sequence analysis of a protein, the substrate is first hydrolyzed selectively with the use of a proteolytic enzyme as catalyst or by the application of an appropriate chemical method. The resulting mixture of peptides usually contains a large number of components that vary in size and differ in composition. However, each peptide must be isolated as a highly purified sample before sequence studies are undertaken.

Peptides are purified by a combination of different methods until each peptide is obtained as a pure species; it is not at all uncommon to use three or more purification procedures. A number of criteria are commonly employed as tests of purity. Paper chromatography in several solvent systems and paper electrophoresis at one or more pH values serve admirably as tests of purity because of their high resolving power, speed of operation, great sensitivity, and because many samples can be tested at one time. Amino acid analysis is another test of purity that is widely used, for the ratio of amino acids must be whole numbers in a pure peptide.

The sequence of purified peptides is deduced from the use of one or more degradative procedures. More than one approach is often required, for degradative techniques sometimes fail. But a multiple approach is also desirable because any overlap of information increases the reliability of the determinations.

Quantitative amino acid analysis of peptides plays an essential role in sequence studies. It establishes the amino acid composition of the intact peptide and identifies those amino acids that are removed sequentially by chemical- and enzymic-degradative procedures. Furthermore, the overall yield can be estimated from the recovery of the sample where the amount of sample is also determined from amino acid analysis.

The group of peptides obtained from the degradation of the protein must account for the total structure of the protein. The amino acid composition of the individual peptides is tallied and the sum is compared with the known composition of the protein. In this way it becomes immediately obvious if one or more peptides are still missing. If the analyses are satisfactory, than the "bookkeeping" arrangement just described helps one to judge if the sequence is complete.

The sequence of peptides derived from one method of degradation is insufficient to describe the complete sequence of the protein. In order to correctly align the peptides and be equivalent to their order in the protein, it is necessary to study the sequence of a second distinctive mixture of peptides. The protein sample is again degraded to a mixture of peptides, each peptide is purified, and the sequences are established.

The method of overlapping sequences of two groups of peptides and arriving at a complete structure is illustrated by the example given in Fig. 5-1. A decapeptide is used for illustrative purposes, but the considerations that apply here

Peptide Mixture *A* Ala•Phe + Gly•Lys•Asn•Tyr + Arg•Trp + His•Val

↑ Method *A*

Decapeptide (Ala, Phe, Gly, Lys, Asn, Tyr, Arg, Trp, His, Val)

↓ Method *B*

Peptide Mixture *B* Ala•Phe•Gly•Lys + Asn•Tyr•Arg + Trp•His•Val

Fig. 5-1. A diagramatic representation of the strategy underlying sequence analysis. A decapeptide represents an unknown peptide (or protein) with its composition shown in brackets. The decapeptide is degraded to a mixture of smaller peptides on hydrolysis with chymotrypsin (Method A). The sequences of the four purified peptides are established. The decapeptide is also degraded to a different mixture of peptides on hydrolysis with trypsin (Method B). The sequences of these three peptides are also determined. The peptides from the A and B mixtures are related to one another by seeking common sequences or overlaps (Table 5-1).

are equally true for large polypeptides, such as proteins. The amino acid composition of the decapeptide is given on the central line with each amino acid residue separated from the next by a comma and the unknown sequence enclosed in a bracket. The decapeptide is degraded in a specific manner (Method A) with the formation of three dipeptides and a single tripeptide (Peptide Mixture A). The four peptides are separated one from another, and the sequence of each is established. [The convention used for a completed sequence is to show the residues connected by dots (or dashes).] A second set of peptides (Peptide Mixture B) is derived from the specific cleavage (Method B) of a sample of the intact decapeptide. These peptides are also purified and the sequence of each must be established.

PEPTIDE OVERLAPPING

The ordering of peptides is based on the occurrence of the same partial sequence in a peptide from Mixture A with one from Mixture B. For example, the dipeptide Ala·Phe in Mixture A is part of the sequence of the tetrapeptide Ala·Phe·Gly·Lys present in Mixture B. The partial overlap extends the sequence of the dipeptide to Ala·Phe·Gly·Lys. Now one looks for a peptide in Mixture A that begins with the sequence Gly·Lys. This is present in the tetrapeptide with the sequence Gly·Lys·Asn·Tyr. The new element in the sequence is Asn·Tyr, and it directs one to the selection of the tripeptide of Mixture B. A summary of the alignment of the peptides with their overlapping sequence is given in Table 5-1. All peptide fragments are completely accounted for in the alignments and lead to the assignment of a unique sequence for the decapeptide.

TABLE 5-1

Peptide Overlaps and Development of an Extended Sequence

Peptide	*Partial Sequence*
A–1	Ala·Phe
B–1	Ala·Phe·Gly·Lys
A–2	Gly·Lys·Asn·Tyr
B–2	Asn·Tyr·Arg
A–3	Arg·Trp
B–3	Trp·His·Val
A–4	His·Val
Sequence of decapeptide	Ala·Phe·Gly·Lys·Asn·Tyr·Arg·Trp·His·Val

ASSIGNMENT OF DISULFIDE BONDS

If a protein contains disulfide bonds or sulfhydryl groups, the sulfur atoms must be converted to stable derivatives before starting the sequence analysis in order to prevent these chemically reactive groups from undergoing undesirable side reactions. The disulfide bonds are cleaved and converted to modified groups. On completion of the amino-acid-sequence determination of the modified protein, the assignment of the disulfide linkages must still be made. Consequently, the protein molecule with intact disulfides is degraded to a mixture of peptides, and those peptides linked by disulfides must be characterized and identified.

The sequence assignment of a protein rests on the acquisition of a large mass of data from repeated degradations. Much of the data acquired as the work progresses serves to reinforce earlier interpretations as well as to enhance the reliability of the sequence. In the final analysis, the amino acid sequence serves as an excellent accounting of the experimental data; but, nevertheless, the sequence itself must be considered as a suitable working hypothesis for the structure of the molecule. With recent accomplishments in organic-chemical synthesis of large polypeptide chains with biological properties, the synthesis of most protein molecules will shortly become possible, thereby providing an additional test of the sequence analysis of the molecule.

REFERENCES

Books

C. H. W. Hirs and S. N. Timasheff (Eds.), "Enzyme Structure," *Methods in Enzymology,* Vol. XXV, Part B. Academic Press, New York, 1972.

S. Blackburn, *Protein Sequence Determination. Methods and Techniques.* Marcel Dekker, Inc., New York, 1970.

S. B. Needleman (Ed.), *Protein Sequence Determination. A Sourcebook of Methods and Techniques.* Springer-Verlag, New York, 1970.

W. A. Schroeder, *The Primary Structure of Proteins.* Harper & Row, New York, 1968.

D. T. Elmore, *Peptides and Proteins.* Cambridge University Press, New York, 1968.

J. L. Bailey, *Techniques in Protein Chemistry*, 2nd ed. Elsevier, Amsterdam, 1967.

Reviews and Articles

A. D. Goldstone and S. B. Needleman, "Reconstruction of the Primary Sequence of a Protein from Peptides of Known Sequence," in *Protein Sequence Determination*, S. B. Needleman (Ed.). Springer-Verlag, New York, 1970.

B. S. Hartley, "Strategy and Tactics in Protein Chemistry," *Biochem. J.,* **119**, 805 (1970).

Modification of Disulfide Bonds

SIX

Proteins that normally function in an extracellular environment generally contain one or more disulfide bonds. Typical examples of these proteins include the immunoglobulins, protein hormones, serum albumin, and the digestive enzymes—trypsin, pepsin, and ribonuclease.

THE DISULFIDE BOND

The disulfide bond forms a bridge between two half-cystine residues, bringing two distant regions of the polypeptide chain into close contact. As a structural element, the disulfide bond contributes to the stability of the folded molecule. A number of residues may intervene between the members of the bridge, and polypeptide loops of different size are observed in the primary amino acid sequences (Fig. 6-1). In insulin and the immunoglobulins, both intermolecular and intramolecular disulfides are present. It should be noted that the intermolecular disulfide represents one of the few examples of a covalent linkage between polypeptide chains in multichain proteins.

Proteins with their disulfides intact cause added complications when submitted to sequence analysis. For example, denaturation of the protein is incomplete, and side chain interactions may still be important in the regions near the disulfides. Consequently, the protein substrate is difficult to hydrolyze with proteolytic enzymes and incomplete hydrolysis is usually observed.

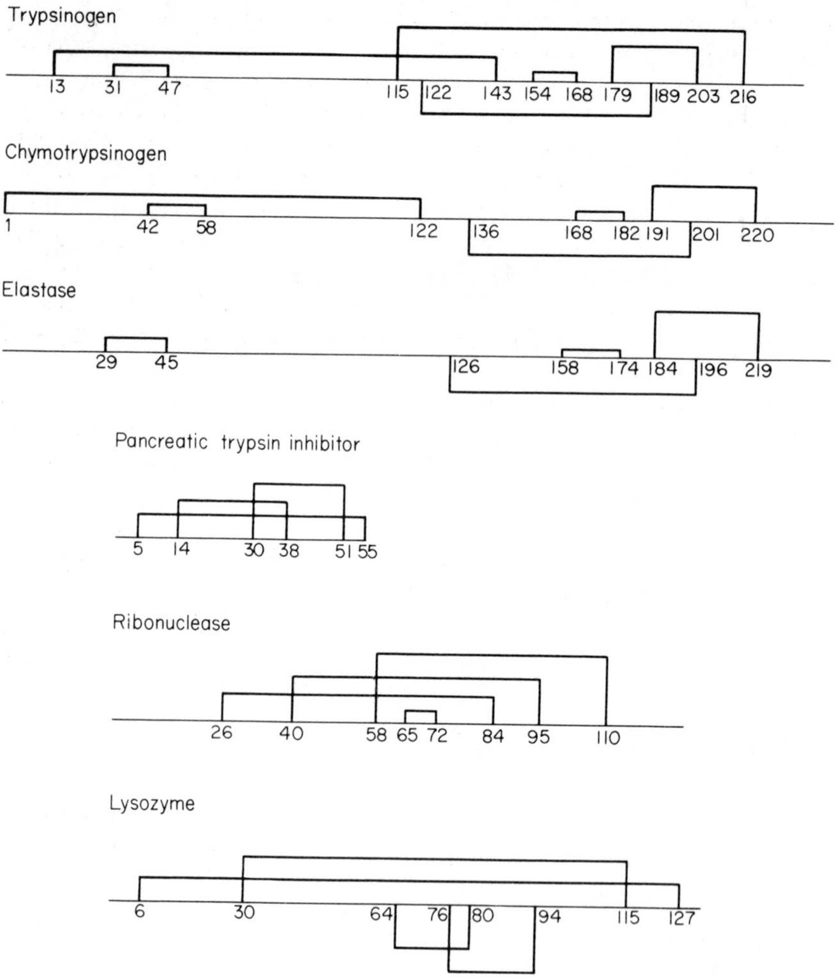

Fig. 6-1. A schematic representation of polypeptide chains with the positions of the disulfide bridges correctly positioned along the chain. The horizontal line represents the polypeptide chain, and its length corresponds to the total number of residues in the protein.

DISULFIDE INTERCHANGE

Furthermore, if several disulfides are present in the molecule, they may participate in an interchange reaction, as first described by Sanger. In Eq. (6-1),

$$R_1{-}S{-}S{-}R_2 + R_3{-}S{-}S{-}R_4 \rightleftharpoons R_1{-}S{-}S{-}R_3 + R_2{-}S{-}S{-}R_4 \qquad (6\text{-}1)$$

the interchange between two disulfides is shown, where R_1, R_2, R_3 and R_4 represent specific amino acid sequences associated with the half-cystine residues. The disulfide interchange creates two difficulties: (a) the correct assignment of the disulfide bridges as they occur in the structure of the protein is obscured by the scrambling effect, and (b) the yield of each peptide containing a disulfide bond is low and the problem of isolating pure peptides becomes more difficult.

The interchange reaction is catalyzed by ionized sulfhydryl groups, and the extent of interchange becomes greater as the pH is raised toward the alkaline region. Knowledge of the reaction mechanism helps one to choose conditions that minimize or eliminate the exchange process; pH values less than 7 must be used and free sulfhydryl groups of the protein should be protected as alkylated derivatives (see below).

For a complete sequence determination of a protein containing disulfide bonds, chemical methods are used to cleave the disulfides and to convert the groups to stable derivatives. In this way the difficulties associated with intact disulfides are circumvented. We will now turn our attention to the modification of disulfide bonds and sulfhydryl groups as a necessary first step in the sequence analysis of proteins. In Chap. 10 we will deal with the specific problem of assigning the disulfide bridges as they occur in the intact protein.

PERFORMIC ACID OXIDATION

Sanger introduced the use of the oxidizing agent performic acid for the modification of protein disulfide bonds. The reagent is prepared by mixing hydrogen peroxide with formic acid, and the oxidation of the protein is carried out at 0°C for 2 hours. Disulfides are quantitatively converted to sulfonic-acid groups, and the oxidized half-cysteine amino acid is called cysteic acid [Eq. (6-2)]. If the protein contains cysteine residues as well, these residues are also

$$\begin{array}{ccc} \mathrm{CH_2{-}S{-}S{-}CH_2} & & \left(\begin{array}{c}\mathrm{CH_2{-}SO_3H} \\ | \\ \mathrm{-NHCHCO-}\end{array}\right) \\ \mathrm{|\qquad\qquad\qquad|} & \xrightarrow{\mathrm{HCOOOH}} \; 2 & \\ \mathrm{-NHCHCO-\qquad -NHCHCO-} & & \end{array} \qquad (6\text{-}2)$$

Cystine residue Cysteic acid residues

oxidized to cysteic acid. Performic acid also raises the oxidation level of the sulfur atom of methionine to the sulfone, and tryptophan is converted to a mixture of oxidized products.

Amino acid analysis of oxidized proteins

Performic acid oxidation is very easy to perform, the protein is recovered free of oxidizing agent, and the extent of oxidation is readily established by amino acid analysis which gives precise values of the cysteic acid and methionine-sulfone content. In fact, the most reliable analytical determination of cystine, cysteine, and methionine is obtained by analysis of performic acid-oxidized samples. However, if cystine and cysteine are both present, an extra determination is required to obtain the cysteine content. This step may be accomplished by the use of one of several excellent colorimetric procedures for the direct estimation of the sulfhydryl content of a protein, or by amino acid analysis of the *S*-alkylated protein (see below).

Application to insulin

In his studies on insulin, Sanger used performic acid to cleave the disulfide bonds. The oxidized sample contained two polypeptide chains: the A chain of 21 residues containing 4 cysteic acid residues, and the B chain of 30 residues with 2 cysteic acid residues. The cysteic acid content suggested that the two chains of insulin are connected by two intermolecular disulfides and that the A chain has a third disulfide as an intramolecular bridge. The two oxidized chains were purified and the sequence of each was established.

It should be pointed out that performic acid oxidation of insulin offered several advantages for the subsequent sequence analysis.

1. Cleavage of the disulfides disrupted the organized structure of the molecule and produced random-chains.

2. The sequence analysis of insulin was performed on the separate polypeptides, each smaller than the intact molecule, and the use of the fragments for sequence analysis made the task much simpler.

3. Cysteic acid residues were stable to all experimental conditions used in sequence analysis.

4. The absence of both tryptophan and methionine provided the simplest case possible for the use of performic-acid oxidation, particularly since the several oxidation products arising from tryptophan were avoided.

If the protein lacks tryptophan, performic acid oxidation is a highly desirable method. However, the multiple products formed from oxidized tryptophan residues increase the complexity of the system and add further difficulties to the sequence analysis.

REDUCTION AND ALKYLATION

Reduction with mercaptoethanol

Protein disulfides are converted to sulfhydryl groups on treatment with a reducing agent. At high concentrations of mercaptoethanol, and at pH 8 to 9, the disulfide bonds are reduced quantitatively in a matter of hours at room temperature. It is necessary to include 8 *M* urea or 6 *M* guanidine hydrochloride to denature the protein, unfolding the polypeptide chain to a random configuration, so that all disulfides become accessible to the reagent. The reaction is readily reversible, as seen from Eq. (6-3), since equivalent type of molecules are

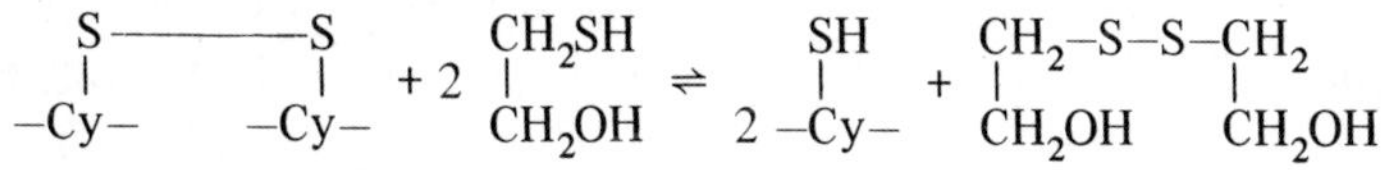

Disulfide bridge　Mercapto-ethanol　Reduced product　Oxidized reagent　(6-3)

both reactants and products. High concentrations of mercaptoethanol (0.1 to 0.5 *M*) are required to drive the reaction to completion.

Reduction with Cleland's reagents

Cleland showed that dithioerythritol and dithiothritol are superior reagents, of low oxidation-reduction potential, capable of reducing protein disulfides at concentrations of the reagent (0.01 *M*) far below that required with mercaptoethanol. Experimental conditions are similar to those used with mercaptoethanol, although a denaturing agent may be avoided with many globular proteins. An examination of the reaction sequence [Eq. (6-4)] shows that the thiol group of

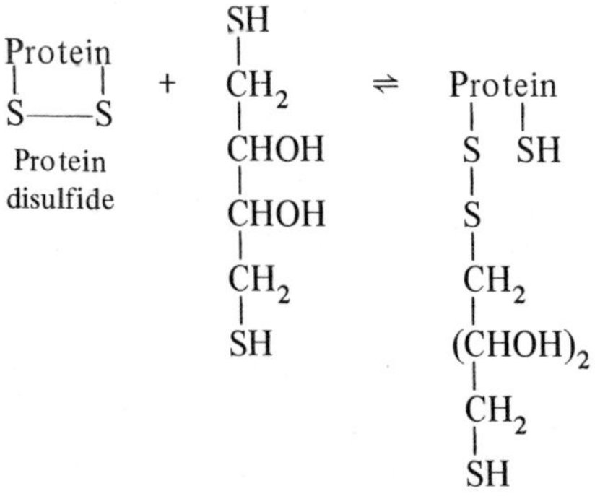

(6-4a)

Protein disulfide　Cleland's reagent　Mixed disulfide

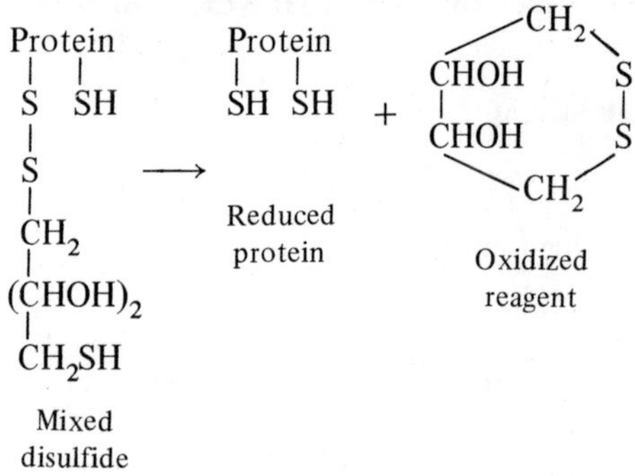

(6-4b)

the reagent attacks a protein disulfide, producing a mixed disulfide as an intermediate [Eq. (6-4a)]. Up to this point the progress of the reaction is essentially the same as it is with mercaptoethanol. However, the reaction goes to completion with the formation of a stable six-membered ring of the oxidized form of the reagent that accompanies the production of the reduced protein [Eq. (6-4b)].

Reaction with iodoacetate and iodoacetamide

The reduced protein is not stable, since sulfhydryl groups are easily oxidized, even by atmospheric oxygen, thus causing the regeneration of disulfide bonds. In order to modify the protein into a form suitable for sequence studies, the reduced protein is treated with an alkylating agent that converts the sulfhydryl group to the stable thioether derivative. For example, reaction with iodoacetate produces *S*-carboxymethylcysteine, [Eq. (6-5)], a derivative that is stable and present

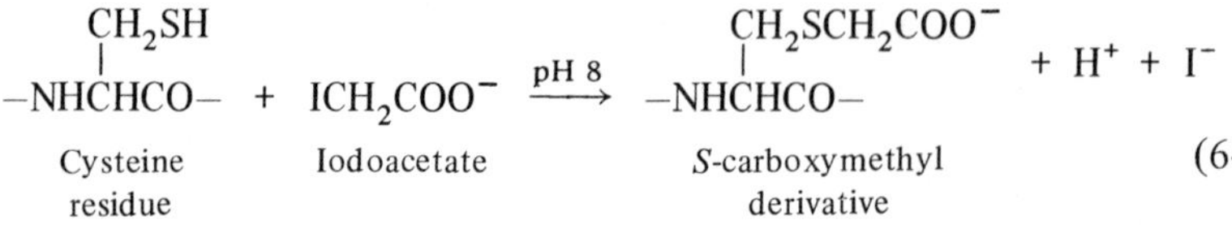

(6-5)

as the anionic specie at neutral pH. Iodoacetamide also reacts in the same way, and the corresponding *S*-carboxyamidomethyl derivative is formed, which is uncharged at neutral pH, a useful property in certain studies.

In order to completely reduce the disulfides of proteins, urea or guanidinium chloride is usually added to denature the protein. The removal of the denaturant is not necessary for alkylation to occur. One simply adds iodoacetate in sufficient amount to react with the reducing agent and with the protein sulfhydryl groups.

The *S*-carboxymethyl derivative is stable to acid hydrolysis, as is generally the case with thioether linkages. On the other hand, the *S*-carboxyamidomethyl derivative is converted on hydrolysis to *S*-carboxymethylcysteine. Amino acid analysis of the modified protein can be used to evaluate the extent of reduction and alkylation. Analysis of the *S*-carboxymethylcysteine derivative also gives the total disulfide and sulfhydryl content of the protein. In order to evaluate the sulfhydryl content, however, the denatured protein must be first treated with iodoacetate, in the absence of a reducing agent, and amino acid analysis of the product gives the *S*-carboxymethylcysteine content, a value that will be equal to the number of residues of cysteine. Thus, if the total *S*-carboxymethylcysteine and that derived from cysteine residues are known, the difference between the two numbers is equal to the disulfide content.

The reaction of iodoacetate with sulfhydryl groups is faster than that of bromoacetate, which, in turn, is very much faster reacting than the chloro derivative. Iodoacetamide reacts more rapidly than iodoacetate, and the rate is greater at pH 8 (the preferred range is pH 7.5 to 9.0) than at lower pH values; the reactive species is the sulfhydryl anion.

Iodoacetate reacts preferentially with sulfhydryl groups, but the reagent also reacts with histidine, methionine, and lysine under other experimental conditions. At pH 5 and above, histidine is alkylated but the reaction time is 24 hours or longer compared with the usual 15 to 30 minutes required for the sulfhydryl group. Above pH 9, the ϵ-amino group of lysine is alkylated. Methionine is alkylated equally well at all pH values but with an extremely slow rate and with reaction times of 2 days or longer.

Alkylation of the reduced protein is restricted to sulfhydryl groups if the pH is close to 8, the reaction time is a few hours or less, and the iodoacetate is added in stoichiometric amounts, or in slight excess, over the sulfhydryl content. Amino acid analysis of the alkylated protein serves as a check of the specificity of the reaction, for the carboxymethyl derivatives of histidine, lysine, and methionine are resolved on ion-exchange chromatography and their absence would clearly indicate that the reaction was restricted to sulfhydryl groups.

Aminoethylation

Sulfhydryl groups are converted to *S*-β-aminoethylcysteine on treatment of the sample with ethyleneimine at pH 8.5 [Eq. (6-6)]. The reaction proceeds

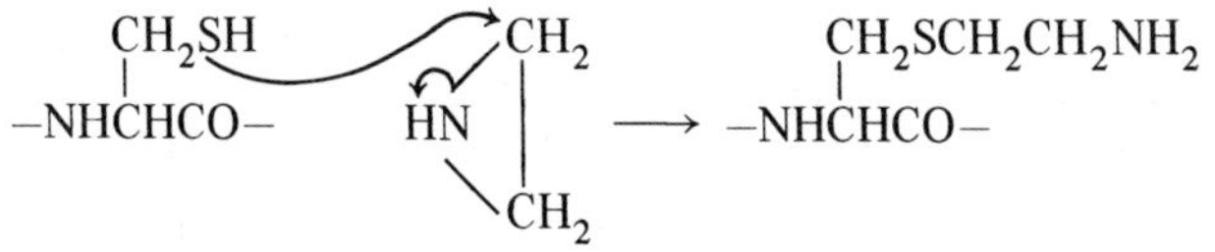

(6-6)

Cysteine residue — Ethyleneimine — *S*-β-aminoethylcysteine by a nucleophilic attack of the sulfur atom at a methylene carbon of the three-membered ring compound. The electron shift to the nitrogen of the amine generates the primary amino group in the terminal position. Under the usual reaction conditions, ethyleneimine is specific in its reaction with sulfhydryl groups.

Aminoethylation is conveniently carried out on the fully reduced protein in the presence of a denaturing agent. The derivative is stable to hydrolytic conditions and the product is estimated on amino acid analysis.

An attractive feature of the aminoethylation reaction described by Eq. (6-6) is the conversion of a cysteine residue into a structure isosteric with lysine.

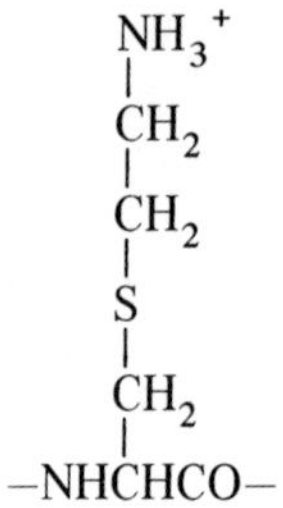

S-β-aminoethylcysteine

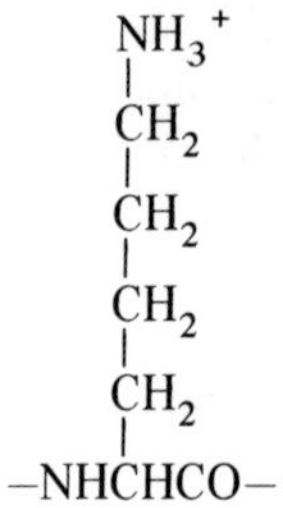

Lysine

The sulfur atom and the methylene group ($-CH_2-$) have almost the same bond length, which places the positively charged amino function in the two structures at equal distances from the peptide bond. As a result, the bond on the carboxyl side of *S*-β-aminoethylcysteine residues can be hydrolyzed by trypsin in the same way as those involving lysyl residues. The introduction of *S*-β-aminoethyl derivatives increases the number of potential points of cleavage of the polypeptide chain—a useful feature for structural studies.

REFERENCES

Books

G. E. Means and R. E. Feeney, *Chemical Modification of Proteins.* Holden-Day, San Francisco, 1971.

C. H. W. Hirs (Ed.), "Enzyme Structure," *Methods in Enzymology*, Vol. XI. Academic Press, New York, 1967.

Reviews and Articles

H. A. McKenzie and D. C. Shaw, "Alternate Positions for the Sulphydryl Group in β-Lactoglobulin: the Significance for Sulphydryl Location in Other Proteins," *Nature New Biology*, **238**, 147 (1972).

W. Konigsberg, "Reduction of Disulfide Bonds in Proteins with Dithiothreitol," *Methods in Enzymology,* **25**, 185 (1972).

R. Cecil, "Intramolecular Bonds in Proteins. I. The Role of Sulfur in Proteins," *The Proteins*, **1**, 380 (1963).

A. M. Crestfield, S. Moore, and W. H. Stein, "The Preparation and Enzymatic Hydrolysis of Reduced and *S*-Carboxymethylated Proteins," *J. Biol. Chem.* **238**, 622 (1963).

C. H. W. Hirs, "The Oxidation of Ribonuclease with Performic Acid," *J. Biol. Chem.,* **219**, 611 (1956).

Degradation of a Protein to a Mixture of Peptides

SEVEN

The sequence of a protein may be derived from the overlap of at least two groups of peptide sequences (Chap. 5). In order to reach this goal, the protein must be degraded in a specific manner to produce a mixture of relatively small-sized peptides. The yield of individual peptides must be sufficiently high so that purification and further degradation may be performed without undue difficulties. The degradation must avoid the extreme case where amino acids, di-, tri-, and tetrapeptides predominate. If this situation occurs, the separation of the peptides requires an enormous effort. Furthermore, it may be difficult to overlap the peptide sequences.

The protein is usually degraded by methods that are selective for peptide bonds involving specific amino acid residues. In this way a limited and predictable cleavage occurs. Proteolytic enzymes and a few chemical methods are generally employed to produce peptide mixtures suitable for sequence analysis. For example, hydrolysis of a protein with the proteolytic enzyme trypsin is restricted to arginyl and lysyl residues.

PEPTIDE BOND CLEAVAGE WITH PROTEOLYTIC ENZYMES

The choice of a suitable proteolytic enzyme is dictated by the specificity characteristics of the enzyme toward protein substrates. A list of the commonly used enzymes and their specificities is given in Table 7-1. These enzymes are endopeptidases, which act on internal peptide bonds, in contrast to the group

TABLE 7-1

Specificity of Proteolytic Enzymes[a]

Enzyme	*Source*	*Major Sites of Action*	*Minor Sites of Action*
Trypsin	Pancreas	Arg, Lys	
Chymotrypsin A	Pancreas	Tyr, Phe, Trp	Leu, Met, His, Asn, Gln
Elastase	Pancrease	Leu, Ile, Ala	Others
Pepsin	Gastric mucosa	Tyr, Phe, Trp, Met, Leu	Ala, Glu, Asp, others
Papain[b]	Papaya plant	Arg, Lys, Gly	Many others
Thermolysin[b]	*Bacillus thermoproteolyticus*	Leu, Ile, Phe	Val, Tyr
Subtilisin[b]	*Bacillus subtilis*	Aromatic and aliphatic residues	

[a]Adapted from A. White, P. Handler, and E. L. Smith, *Principles of Biochemistry*. McGraw-Hill, New York, 5th ed., 1973.

[b]The use of these proteolytic enzymes is usually restricted to the further hydrolysis of isolated peptides rather than for the initial proteolysis of a protein substrate.

of exopeptidases, such as the carboxypeptidases and aminopeptidases, which catalyze the hydrolysis of peptide bonds of terminal residues. Endopeptidases hydrolyze peptide bonds on the carbonyl side of susceptible amino acid residues. However, the action of pepsin is favored by the presence of aromatic amino acid residues on both sides of the susceptible peptide bond. Thus trypsin may hydrolyze the linkage between Arg-X, and pepsin that of Phe-Phe.

Denaturation of substrates

Native proteins are usually refractory to the action of proteolytic enzymes, simply because the susceptible residues are buried within the molecule or are inaccessible because of side chain interactions in the folded structure. Those proteins containing disulfide bonds would have been denatured previously when they were oxidized with performic acid or reduced and alkylated (Chap. 6). Proteins lacking disulfides may be denatured at elevated temperatures, with 8 *M* urea, 6 *M* guanidine hydrochloride, or with organic solvents. Thermal denaturation takes place when an aqueous solution of a protein is kept at 100 degrees for 5 to 15 minutes resulting in the formation of a copious precipitate. Denaturation with high concentrations of urea of guanidine hydrochloride, or with an organic solvent, must be followed by dialysis, which effectively removes the denaturing agent. Since trypsin is active in 2 *M* urea, the 8 *M* urea solution may be diluted to this final concentration.

Tryptic hydrolysis

Usually trypsin is a good choice for one of the required cleavages, for hydrolysis is limited to peptide bonds of lysine and arginine, residues that are universally found in proteins. Thus, if a protein contains a number of these residues (about 10 percent), proteolysis produces peptides with an average size of 10 residues. All lysyl and arginyl residues are susceptible except those in the sequences Lys·Pro and Arg·Pro, which are resistant because of the presence of an imino bond. Each tryptic peptide terminates in lysine or arginine, with one possible exception, which would be the peptide derived from the carboxyl-terminal sequence of the protein.

Chymotryptic hydrolysis

Chymotrypsin is another proteolytic enzyme widely used for the degradation of a protein substrate. The major points of cleavage are at aromatic residues, and the resulting peptides terminate in either tyrosine, phenylalanine, or tryptophan. Chymotrypsin also hydrolyzes peptide bonds of several other residues (Table 7-1), but the extent of hydrolysis is variable and it depends on neighboring residues. The yield of peptides from the minor points of hydrolysis may be low.

Peptides from a tryptic and chymotryptic hydrolysis are normally sufficient to overlap the related peptides and to arrive at the complete sequence of the protein. In some studies, it may be necessary to use a third method of cleavage to complete the sequence analysis.

Comments

The degradation of a protein with proteolytic enzymes produces some peptides in high yield while others may be low. The variable yields are directly related to the difficulty of hydrolyzing all susceptible bonds quantitatively, since the immediate neighboring sequence usually differ. Consequently, the rates of hydrolysis are rapid with some sequences and much slower for others. Furthermore, as peptides accumulate during the hydrolysis, product inhibition of the enzyme slows the rate. During a lengthy hydrolysis, limited inactivation of the enzyme may occur, which also slows the rate.

Experimental conditions

The difficulties associated with proteolysis cannot be overcome, and it is necessary to perform trial experiments to arrive at the best compromise possible. Generally the enzyme (at a concentration level of 0.05 percent) is added to the substrate (at a final concentration of 0.5 to 1 percent), the pH is maintained at pH optimum for the enzyme, and the temperature is held at 37°C. The time of hydrolysis is arrived at experimentally, where the optimum may be short (2 to 4 hours) for some proteins and long (18 to 24 hours) for others.

The progress of the reaction can be followed by one of several procedures. One procedure makes use of a pH stat, a sophisticated modification of a pH meter capable of maintaining a reaction at a constant pH by the automatic addition of alkali (or acid) as required and simultaneously recording the consumption of alkali (or acid) as a function of the reaction time. With hydrolysis of a peptide bond at pH 8.5, an ionized carboxyl group, an unionized amino group, and a proton are produced. The pH stat adds alkali as needed to balance the proton release, maintaining a constant pH as hydrolysis proceeds. A progress curve provides a convenient way to monitor the hydrolysis continuously. In the early periods of hydrolysis, alkali consumption is rapid, but the rate slows down or stops as the reaction approaches completion (Fig. 7-1). Furthermore, the number of peptide bonds hydrolyzed may be calculated from the alkali consumption and the amount of substrate present, and the experimentally derived value may be compared with the theoretical value, based on the number of susceptible residues present in the molecule.

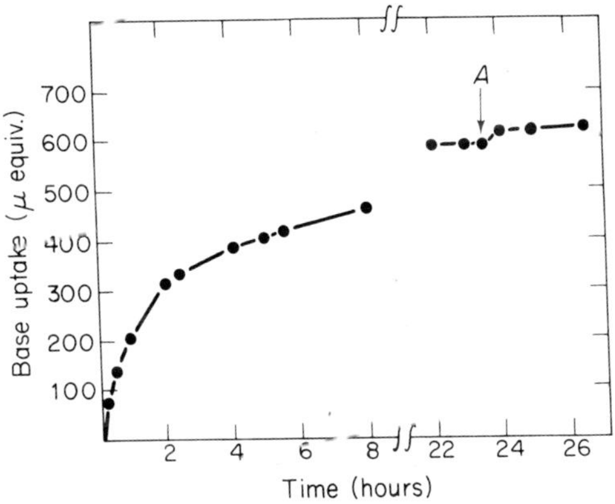

Fig. 7-1. The rate of hydrolysis of oxidized papain (1 percent) by chymotrypsin (0.03 percent) at pH 7.8 and 25 degrees. The base uptake refers to the addition of 0.71 *N* NH_4OH required to maintain constant pH during the digestion. [From A. Light and E. L. Smith, *J. Biol. Chem.*, **235**, 3144 (1960).]

The number of peptide bonds hydrolyzed also can be estimated from a colorimetric analysis of new amino-termini with the ninhydrin reagent. Ninhydrin, as commonly used for analytical measurements, reacts with both α-amino and ϵ-amino groups of proteins, peptides, and amino acids. The number of α-amino groups increases as hydrolysis continues, but the ϵ-amino groups remain constant. The progress of hydrolysis is followed from the increase in the ninhydrin reaction.

Another way to follow a proteolytic digestion is with the "fingerprint technique," also known as peptide mapping (Fig. 7-2). Before describing the technique, it should be noted that peptide mapping is extensively used to examine the structure of closely related proteins. For example, proteins from different biological species and genetically altered proteins usually show limited alterations in sequence, and these changes are readily detected in the fingerprint patterns (see Chap. 11).

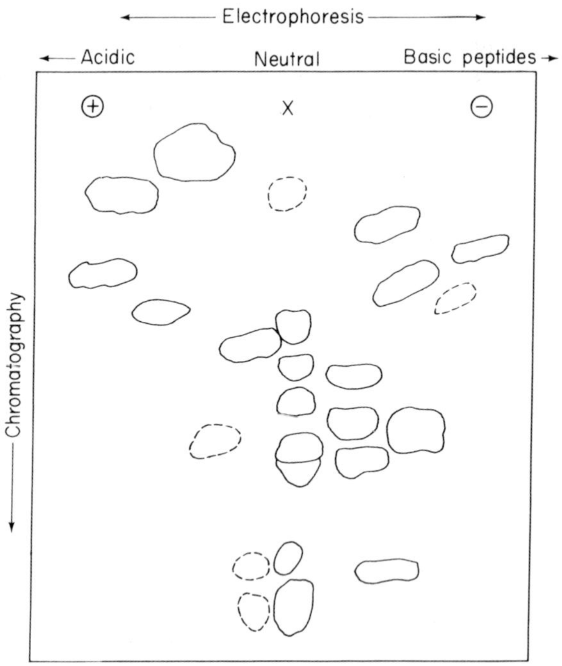

Fig. 7-2. Peptide Mapping. The two-dimensional separation of a peptide mixture. The first dimension represents the separation of peptides by electrophoresis at pH 6.5. The second dimension is run at 90 degrees to the first direction, and the separation is accomplished by chromatography with a solvent system of butanol-acetic acid-pyridine-H_2O. [Adapted from A. Light, B. C. Hardwick, L. M. Hatfield, and D. L. Sondack, *J. Biol. Chem.*, 277, 6289 (1969).]

The peptide mixture is submitted to peptide mapping by initially applying a sample of the digest to a large sheet of filter paper. The sample is submitted to a two-dimensional separation: the first is electrophoresis at pH 6.5, or some other suitable pH, and the second dimension is a separation with a chromatographic-solvent system. The paper sheet is treated with the ninhydrin reagent and possibly also with specific reagents for particular amino acids.

The peptide mixture is usually well resolved, and the number of peptides that one sees can be related to the predicted number based on the composition of the substrate and the known specificity of the proteolytic enzyme. A rough estimate may be made of the yields of individual peptides from the intensity of their color with ninhydrin compared to the color yield of known samples.

In the early stages of proteolysis, hydrolysis is incomplete, and the peptide map shows a greater number of peptides than expected. Many peptides are present in low yield, and their color with ninhydrin is very weak. As proteolysis continues, the number of peptides decreases and their yields increase. The digestion is considered complete when a peptide map is identical to one at an earlier time, the number of peptides is close to theoretical, and the color intensities are all reasonbly strong.

The peptide map reveals the net charge of each of the separated peptides, particularly when the electrophoretic step is performed at pH 6.5 (Fig. 7-2). Those peptides that are electrically neutral at pH 6.5 lie on a line that starts at the origin and falls perpendicular to the direction of electrophoresis. Acidic and basic peptides, with a net charge of one, lie on either side of the neutrals and about equally far away. Peptides with a net charge of 2 or greater lie on lines still farther removed from the origin. Although these details are not required in order to evaluate a proteolytic digestion, peptide maps offer invaluable information that serves as a guide to the purification of the peptide mixture and later, for the testing of isolated peptides for purity (Chap. 8).

Tryptic hydrolysis of *S*-β-aminoethyl-cysteine residues

For several specific applications in sequence studies, it is possible to alter the course of the tryptic hydrolysis of a protein and to obtain hydrolysis at other than lysine and arginine residues. An increase in the total number of susceptible residues occurs when cysteine residues of a protein are converted to the *S*-β-aminoethyl derivative [see Eq. (6-6)]. It should be recalled from an earlier discussion that *S*-β-aminoethylcysteine is isosteric with lysine; the sulfur atom is approximately equivalent to the dimensions of the carbon atom of a methylene group. Thus the primary amino group of *S*-β-aminoethylcysteine is about the same distance from the peptide backbone as in normally the case with the ε-amino group of lysine. Consequently, the *S*-β-aminoethyl derivative meets the specificity requirements of trypsin and it is readily hydrolyzed by the enzyme.

Proteins with cysteine residues are converted directly to the *S*-β-aminoethyl derivative on reaction with ethyleneimine. If cysteine is absent but disulfides are present, reduction of the disulfides, and then alkylation with ethyleneimine, leads to an increase in the number of *S*-β-aminoethyl derivatives. For those proteins with low contents of arginine and lysine, the additional hydrolysis possible at *S*-β-aminoethylcysteine may make it easier to obtain a good distribution of peptides from all regions of the molecule. Since the sequence at the disulfide bridges may in itself be important in comparative structural studies, the tryptic hydrolysis at *S*-β-aminoethylcysteine residues may give peptide fragments suitable for further study of these parts of the molecule.

Acylation of amino groups

Tryptic digestion is restricted to the hydrolysis of arginine residues when the ϵ-amino group of lysine is converted to a derivative that masks the amino function. Suitable blocking groups lengthen the side chain, change the charge to a neutral or acidic function, and alter the ability of the modified lysine residues to meet the specificity requirements of trypsin.

The choice of a suitable reagent is dictated by the specificity of its reaction toward lysine residues, by the availability of suitable analytical procedures to follow the progress of the reaction, by the need to obtain a quantitative conversion to products, and by the stability of the protein during the course of the reaction. Acetic anhydride, which has long been used as a reagent for the modification of proteins, has a specificity directed toward the ϵ-amino group of lysine. The reagent also acetylates tyrosine phenolic groups, but this amino acid is regenerated at the slightly alkaline reaction conditions generally used; or the *O*-acetyl derivative may be hydrolyzed rapidly on treatment with hydroxylamine at alkaline conditions. Succinic anhydride reacts with protein ϵ-amino groups in a similar manner. The *N*-succinyl derivative is anionic, whereas the *N*-acetyl derivative is neutral, and both change the net charge of the protein.

The *N*-acetyl and *N*-succinyl derivatives of the primary amino groups are stable and are difficult to remove. Although these modifications redirect tryptic digestion, *N*-acylated derivatives cause difficulties in sequence studies (Chaps. 8, 9). For example, peptide purification may be more difficult; the blocked group prevents further degradation of large peptides by trypsin; sequence analysis with the Edman procedure and the use of proteolytic enzymes becomes more difficult or they may fail to work. For these and other reasons, it is preferable to employ a reversible blocking group.

Maleylation

Fortunately, reagents that form easily reversible derivatives with amino groups are available. We will restrict the discussion to the use of maleic anhydride and other substituted maleyl reagents (see Chap. 10). The reaction of maleic anhydride with the amino group is shown in Eq. (7-1), a reaction that is similar to those discussed earlier for acetic and succinic anhydrides. The maleyl derivative increases the ultraviolet absorption of a protein at 250 nm. An estimate can be

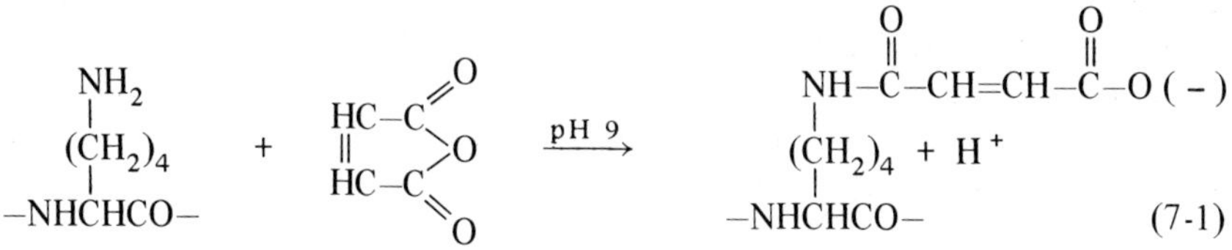

(7-1)

made of the extent of the reaction from the change in the absorption of the protein.

The maleyl derivative is stable at neutral pH and under the conditions of hydrolysis with trypsin. Tryptic hydrolysis of modified substrates is directed to the cleavage at arginyl residues only; and *N*-maleyl tryptic peptides terminate in arginine, as expected. Blocked lysines at internal positions provide potential points for further tryptic cleavage. The lysine derivative is deblocked when the sample is acidified to pH 3.5 and held at 37 degrees for 20 hours (the half-life of the reaction is 10 hours). A further treatment of the peptide with trypsin will result in the hydrolysis of the regenerated lysyl residues producing two or more peptides. The maleylation procedure is illustrated by the example given in Fig. 7-3.

Maleylated proteins are also useful for other purposes. The protein derivatives are often water soluble under conditions where the parent molecules may be insoluble, which make them convenient for protein modification reactions and for molecular weight determinations of insoluble molecules. In the latter case, measurements in the ultracentrifuge or with calibrated dextran gel columns can be carried out with aqueous solutions, thereby avoiding the use of detergents, urea, or guanidine. Not only do maleylated proteins show satisfactory solubility characteristics, but they also remain in solution in a monomeric state, thus eliminating problems associated with aggregation. Indeed, this property has been valuable in the study of proteins with multiple polypeptide chains. The subunits readily dissociate as the maleyl derivatives, which simplifies their separation and further characterization.

The mechanism for deblocking maleyl derivatives is thought to involve an intramolecular catalysis, with the neighboring carboxyl group participating in the hydrolysis (Fig. 7-4). Intramolecular ring closure leads to displacement of maleic anhydride, which is hydrolyzed in solution to maleic acid. If 0-maleyl tyrosine is formed, it would be expected to decompose immediately at the alkaline pH of the reaction [Fig. 7-4(b)].

A reagent for the reversible blocking of lysyl residues with an easily removable derivative is citraconic anhydride (2-methylmaleic anhydride). Deacylation of the citraconyl derivative is accomplished rapidly at 20 degrees and pH 2 to 3, where the half-life of the reaction is 1 to 2 hours. If acylation is carried out with 2,3-dimethylmaleic anhydride, the resulting derivative is even less stable.

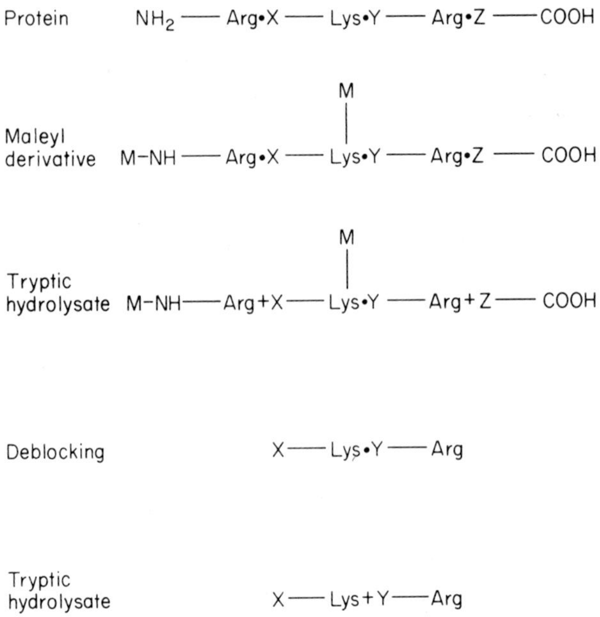

Fig. 7-3. Redirected tryptic hydrolysis of a polypeptide with hydrolysis restricted to arginine residues. The ϵ-amino groups of lysine residues are converted to the maleyl derivative (M). Trypsin catalyzes the hydrolysis at arginine residues. The peptides are separated, and each peptide with a blocked lysine residue is unblocked at pH 3. The peptides containing lysine are again hydrolyzed with trypsin to form smaller fragments.

Derivatives of arginine

Tryptic hydrolysis of a protein or peptide can be redirected to the cleavage of lysine residues only. Arginine residues will no longer be hydrolyzed if the guanidinium group is converted to a stable derivative. The reaction of arginine with 1,2 and 1,3-dicarbonyl compounds gives condensation products that are stable because of the formation of heterocyclic rings. In Eq. (7-2) the reaction is shown with the reagent 1,2-cyclohexanedione. The condensation reaction with malonaldehyde is given in Eq. (7-3).

The chemical modification of arginine residues has principally been used in structure-to-function relationships rather than in sequence studies. The limited use of chemical derivatives of arginine is associated with the drastic reaction conditions required (0.2 *M* NaOH and 10 *N* HCl), difficulties in estimating the extent of reaction, and a possible lack of specificity, particularly toward lysines.

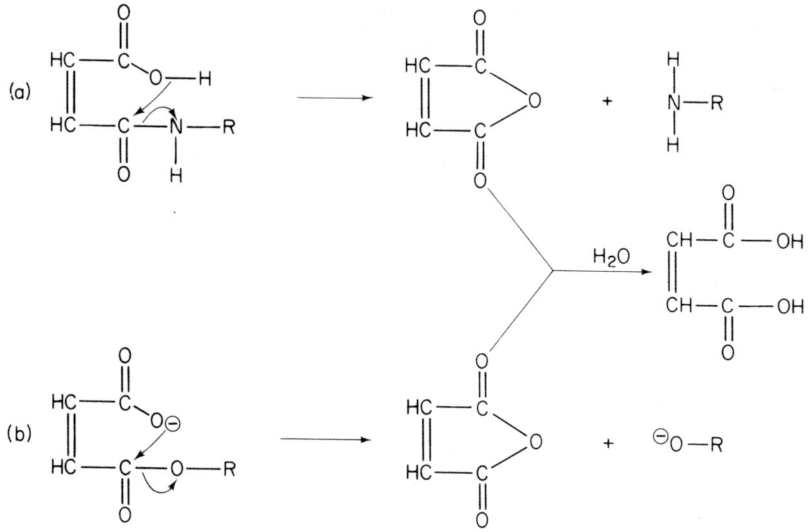

Fig. 7-4. Suggested mechanisms for the hydrolysis of maleyl derivatives catalyzed by the carboxyl group. Reaction (a) shows the hydrolysis of amide bonds at acid pH, and reaction (b) shows the hydrolysis of *0*-maleyl tyrosine at alkaline pH.

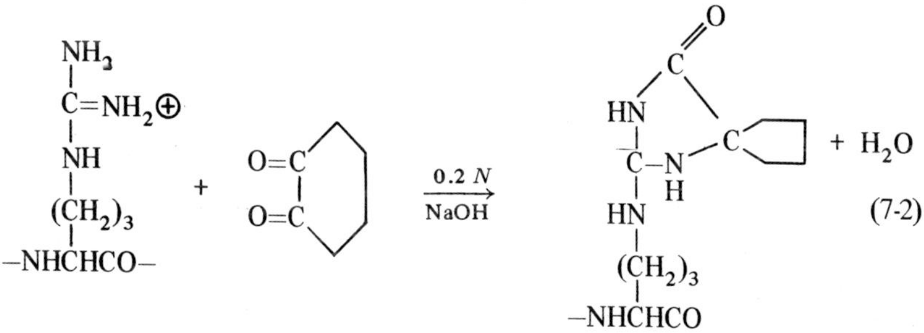

(7-2)

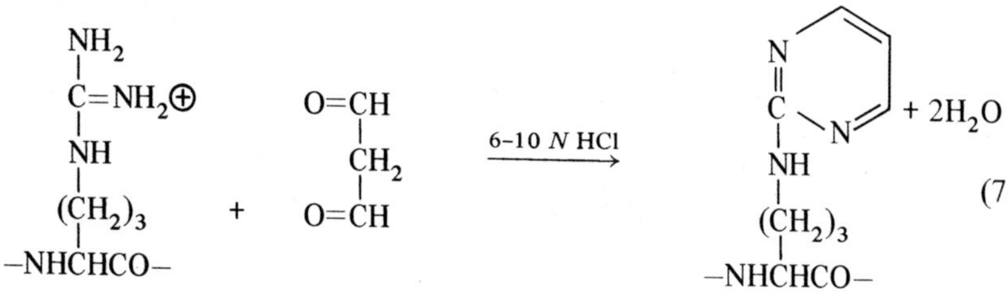

(7-3)

CHEMICAL METHODS

Partial acid hydrolysis

Sanger used partial acid hydrolysis (0.1 *N* HCl at 110° or 6 HCl at 37°) to degrade each of the performic-acid-oxidized insulin chains to small peptides, which were then subjected to sequence analysis. The principal products of acid hydrolysis are tripeptides, dipeptides, and amino acids; and with proteins much larger than insulin, the number of fragments is so great that the separation of the mixture is extremely difficult.

Many attempts have been made to restrict the course of acid hydrolysis and to limit the number of susceptible residues. Hydrolysis is restricted mainly to aspartic acid residues if the experimental conditions are 0.03 *N* HCl, 105°, and the time between 8 to 48 hours. Aspartic acid residues are preferentially cleaved because of intramolecular catalysis (Fig. 7-5). The specificity of the reaction is high for aspartyl residues, but hydrolysis also occurs at much slower rates at serine, threonine, and glycine residues. Since these residues are ubiquitous and normally fairly abundant in proteins, their hydrolysis, to even a small extent when large proteins are studied, increases the complexity of the digestion mixture. Consequently, partial acid hydrolysis is usually restricted to peptides.

Cleavage at methionine residues

The single most useful chemical technique is the selective cleavage of peptide bonds of methionine residues with cyanogen bromide. The reaction mechanism and some of the reaction conditions are given in Fig. 7-6. The initial product is an unstable intermediate, a cyanosulfonium derivative. The elimination of methyl thiocyanate occurs simultaneously with a nucleophilic attack of the carbonyl oxygen at the sulfur atom. The driving force of the cleavage reaction is the formation of the five-membered ring structure. The imino bond is acid-labile, and hydrolysis occurs immediately, cleaving the peptide bond on the carbonyl side of methionine.

The newly-formed carboxyl-terminal residue is homoserine lactone, and it is stable under the usual conditions of sequence analysis. The extent of cyanogen bromide cleavage is estimated from the amount of homoserine lactone produced. This quantity is determined from amino acid analysis, where the homoserine lactone and homoserine produced on hydrolysis are easily resolved on ion-exchange chromatography.

The reaction mechanism suggests that peptide bond cleavage with cyanogen bromide is specific for methionine residues, and indeed this is the case. The cleavage occurs in high yield, and the use of formic acid as solvent serves to denature the protein and expose all methionine residues. Most proteins contain

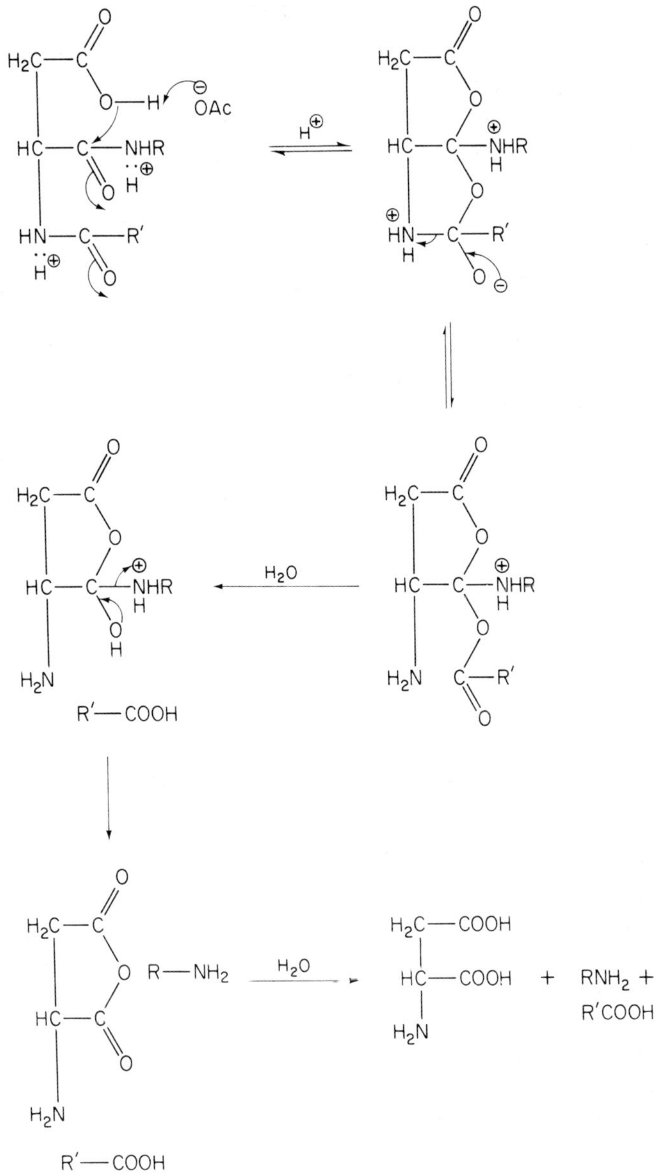

Fig. 7-5. Suggested mechanism for the hydrolysis of the peptide bonds of aspartyl residues.

a limited number of methionine residues, and cyanogen bromide cleavage produces a few large peptides. The highly specific cleavage offers certain advantages in sequence studies.

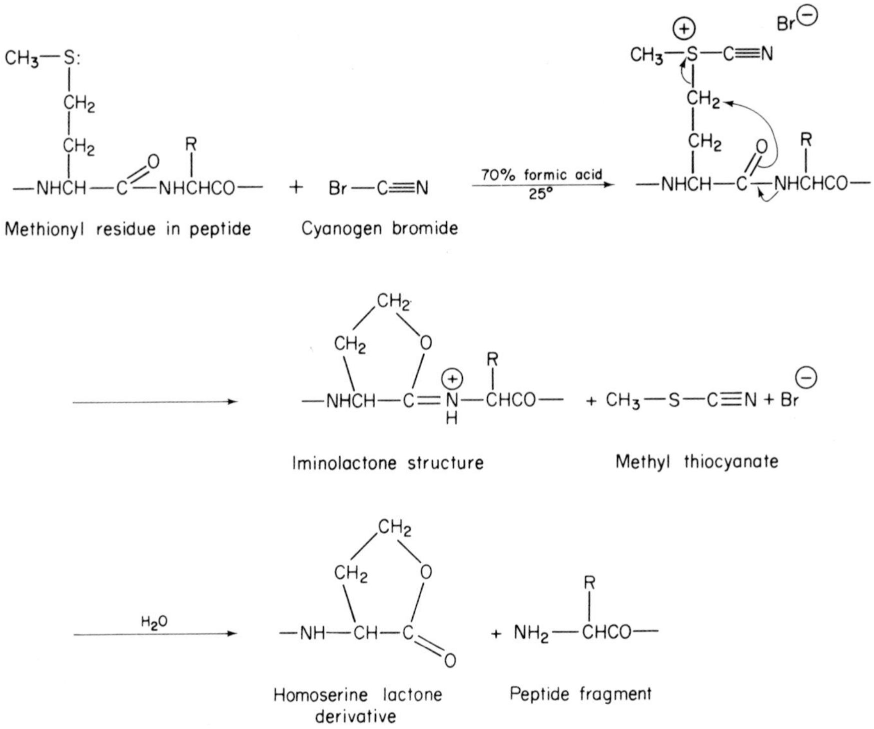

Fig. 7-6. The chemical cleavage of a peptide bond at a methionine residue–*the cyanogen bromide degradation.*

REFERENCES

Books

P. D. Boyer (Ed.), *The Enzymes*, Vol. III, 3rd ed. Academic Press, New York, 1971.

C. H. W. Hirs (Ed.), "Enzyme Structure," in *Methods of Enzymology*, Vol. XI. Academic Press, New York, 1967.

Reviews and Articles

C. B. Kasper, "Fragmentation of Proteins for Sequence Studies and Separation of Peptide Mixtures," in *Protein Sequence Determination*, S. B. Needleman (Ed.). Springer-Verlag, New York, 1970.

T. F. Spande, B. Witkop, Y. Degani, and A. Patchornik, "Selective Cleavage and Modification of Peptides and Proteins," *Advances in Protein Chem.*, **24**, 98 (1970).

R. L. Hill, "Hydrolysis of Proteins," *Advances Protein Chem.*, **20**, 37 (1965).

Separation of Peptides

EIGHT

As many as 20 to 100 peptides may be produced when a protein of average size is hydrolyzed with a proteolytic enzyme. The resolution of a complex mixture of peptides is difficult and time consuming, requiring methods of high resolving power. Even then several procedures may be used before each peptide is purified to a state of homogeneity (see Chap. 5). Fractionation procedures take advantage of differences in size, solubility, and net charge.

GEL FILTRATION

The separation of peptides is accomplished by molecular sieving where differences in particle size contribute to the resolution (Chap. 3). A typical separation of peptides on a column of a dextran gel is shown in Fig. 8-1. The order of elution of components generally reflects their size differences, with the largest molecule first to appear, followed by the next largest, and so forth. The chromatographic separation is followed by direct absorption measurements of the eluent in the ultraviolet at 280 nm (aromatic groups) or 230 nm (peptide bonds), or from the color developed with the ninhydrin reagent.

It is evident from the results shown in Fig. 8-1 that gel filtration appears to provide a limited separation of a peptide mixture, with several fractions overlapping one another. Nevertheless, most peptides will be located in only a few fractions, and further purification of unresolved peptides will be made simpler by this initial step.

A wide selection of solvents is possible with gel filtration systems. Buffered solutions of various types are suitable, and a favorable pH can be selected to

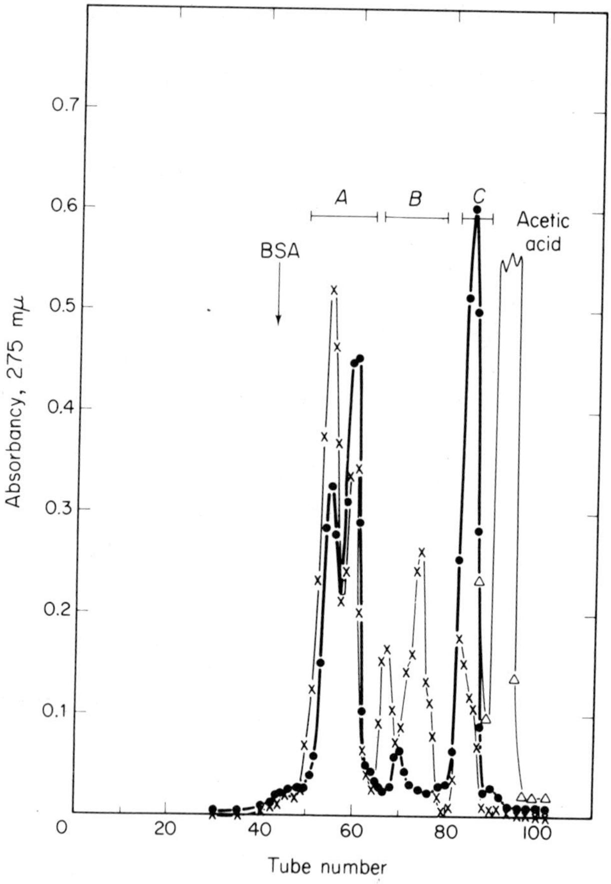

Fig. 8-1. Separation of peptides from a tryptic digest of 20 mg of performic acid-oxidized ribonuclease on a column (310 × 0.9 cm) of Sephadex G-25. The column was operated at 25 degrees with 0.2 *M* acetic acid and 2 ml fractions were collected. The curves with solid circles were obtained from ultraviolet absorption measurements at 275 nm, and the curves with x's were obtained from ninhydrin analyses after alkaline hydrolysis of the samples. The arrow labeled BSA shows where bovine serum albumin emerges from the column. The acetic acid band indicated by triangles was detected by spectrophotometric analysis at 230 nm. [From D. L. Eaker, T. P. King, and L. C. Craig, *Biochem.*, 4, 1479 (1959). Copyright © 1959 by the American Chemical Society. Reprinted by permission of the Copyright owners.]

increase solubility of peptides. One may also use concentrated solutions of acetic acid, formic acid, or urea if insoluble peptides are present in the mixture.

ION-EXCHANGE CHROMATOGRAPHY

Synthetic resins with acidic and basic functional groups, cation and anion exchangers, respectively, are useful chromatographic media for the separation of peptides. Dowex 50 is a commercially available cation exchanger with sulfonic-acid groups (Chap. 4). The resin is stable to extremes of pH and temperature, and it displays a high capacity for peptides.

The resin is initially equilibrated with a 0.2 *M* buffer at pH 3.1, and an acidified solution of a peptide mixture is added to a column containing the resin. Acidic peptides bind poorly and are eluted at pH 3.1 with the 0.2 *M* buffer (Fig. 8-2). Neutral and basic peptides are displaced from the resin with a buffer increasing in both pH and ionic strength, and the peptides are eluted from the column in a sequential order, based largely on their relative cationic properties.

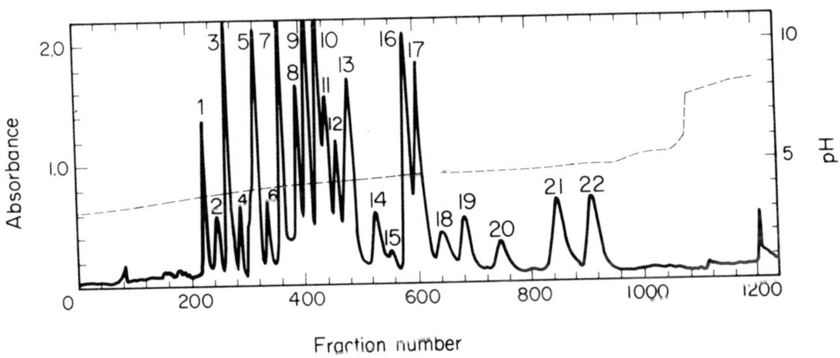

Fig. 8-2. Chromatography of the tryptic peptides of human hemoglobin A on Dowex 50. The ninhydrin color value is plotted against fraction number, and the dashed line indicates the pH of the effluent. The mixture of peptides was resolved on a column (3.5 x 100 cm) of the resin initially equilibrated with 0.2 *N* pyridine acetate buffer at pH 3.1. A buffer gradient was used to increase the pH to 5.0 and the pyridine concentration to 2 *N*. [From W. A. Schroeder, R. T. Jones, J. Cormick, and K. McCalla, *Anal. Chem.*, **34**, 1570 (1962). Copyright © 1962 by the American Chemical Society. Reprinted by permission of the Copyright owners.]

The change in buffer composition is conveniently generated by allowing a buffer of high pH and molarity to flow into a buffer chamber containing the starting buffer. A linear gradient is formed with the apparatus shown in Fig. 8-3. A decided advantage is gained in the study of column fractions when volatile buffers are used, such as pyridine-acetate, for evaporation of the collected fractions provides peptides in a salt-free state.

The progress of the column separation is followed by analyzing aliquots of the fractions with the ninhydrin reagent. Since some peptides may be ninhydrin negative, the colorimetric analysis is also performed on samples hydrolyzed with

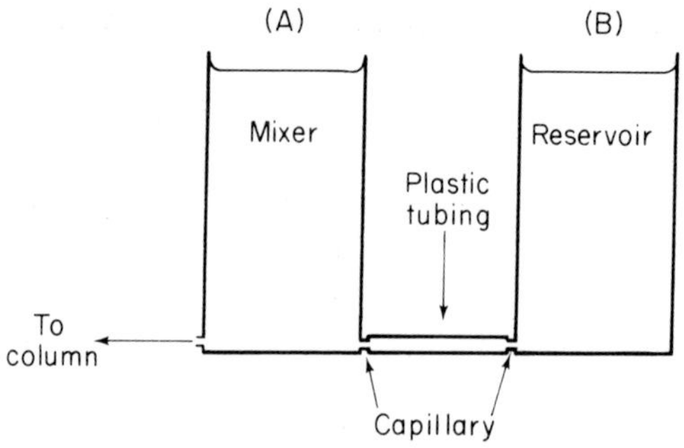

Fig. 8-3. A linear gradient apparatus. Vessel A is stirred with a magnetic stirrer and serves as a mixing container. Solution in vessel B enters vessel A by a capillary tube at a rate equal to the rate of buffer entering the column.

2 *N* NaOH for 2.5 hours at 100 degrees and then neutralized with 30 percent acetic acid, prior to the addition of ninhydrin.

The separation of peptides on Dowex 50 is related to the strength of the electrostatic interactions at a given pH and ionic strength. However, the separation is also influenced by van der Waals interactions of hydrophobic groups with the resin. Indeed, peptides with several hydrophobic groups may bind so strongly that their elution is possible only at an elevated pH and high ionic strengths, if at all. The column separation is normally performed at temperatures of 30 to 50 degrees, which increase the rate of equilibration between the moving phase (buffer) and the stationary phase (peptide bound to the resin).

The resolution of peptides on a column of Dowex 50 is generally quite high with many column fractions containing pure peptides. Those fractions that still contain mixtures of peptides may be rechromatographed on the same resin at a slightly lower pH and ionic strength than that used in their initial separation.

Peptide mixtures can also be fractionated on columns of the resin Dowex 1, a quaternary ammonium ion resin. A typical separation is shown in Fig. 8-4. A buffer of high pH is used initially, and then a pH gradient is started with acetic acid of increasing molarity, which decreases the pH to a lower value. Basic peptides are eluted early, acidics last, and neutral peptides appear between the two groups.

Ion-exchange resins display a high sample capacity, and large columns may be used to separate gram amounts of peptides. The sample load is directly related to the diameter of the column, and the resolving power is dependent on the column height. A hydrolysate of 1 g of protein may be fractionated on a 2.5 $\times$ 150 cm column. The time for the operation of a column varies from 3 to almost 7 days, the range depending on the number of peptides.

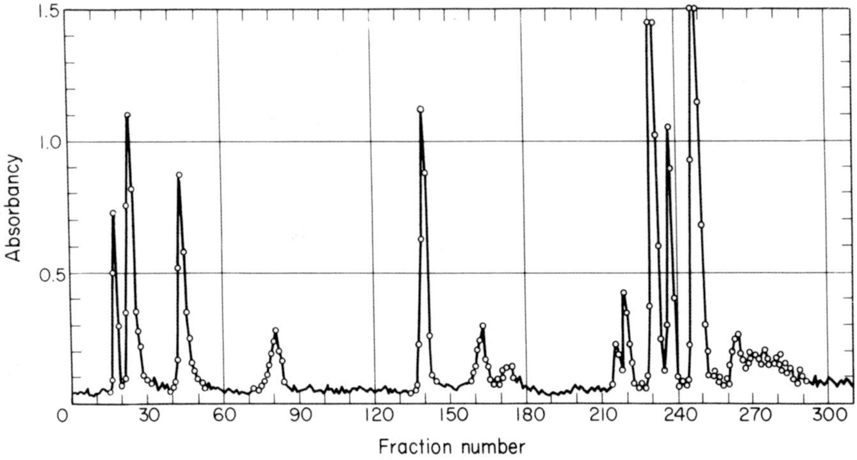

Fig. 8-4. Chromatography of the tryptic peptides of Tobacco Mosaic Virus on Dowex 1. The ninhydrin color value is plotted against fraction number. The mixture of peptides was resolved on a column (2 × 150 cm) of the resin initially equilibrated with a collidine-pyridine buffer of pH 8.3. After 65 fractions, a gradient elution technique was used. A solution of buffer was gradually replaced by 0.1 *N* acetic acid; after fraction 140, the 0.1 *N* acetic acid was replaced by 2 *N* acetic acid; and after fraction 250, by glacial acetic acid. [From H. G. Wittmann and G. Braunitzer, *Virology*, 9, 726 (1959).]

PREPARATIVE PAPER CHROMATOGRAPHY

Paper chromatography of peptides is rapid and simple to perform; it is relatively easy to resolve a number of different samples in a 24-hour period. A peptide mixture is applied along a line parallel to the top edge of a large sheet of filter paper (Fig. 8-5). Then the same end of the paper sheet is dipped into a glass trough containing the organic phase of the solvent system. In most cases, the solvent systems are prepared by equilibrating *n*-butanol with either dilute acetic acid or a solution of acetic acid and pyridine. The solvent travels through the pores of the paper by capillary action, with the aqueous phase held by the paper fibers and the organic phase moving from the origin to the end of the paper. Peptides soluble in the organic phase move from the origin, with their rate of movement directly related to their relative solubility.

After a chromatographic separation is completed, the paper sheet is removed from the chromatography cabinet and the paper is allowed to dry in an airstream until all traces of solvent are removed. The location of peptides on the chromatogram is revealed by treating narrow guide strips—removed from the side edges of the sheet—with the ninhydrin reagent (Fig. 8-5). The peptides in each band are recovered by irrigating the appropriate paper strip with water in an enclosed chamber. At times the recovery of peptide may be low, particularly if it contains several hydrophobic residues, and improved yields may be found on elution with 50 percent acetic acid.

Several micromoles of a peptide mixture can be resolved on a single large sheet of filter paper, thus providing a sufficient amount of pure peptide for amino acid analysis and degradative studies. Since the procedure is simple and fast, and requires very little in the way of apparatus, paper chromatography may be used as the only method of peptide purification, or, more generally, after a preliminary separation with a column system.

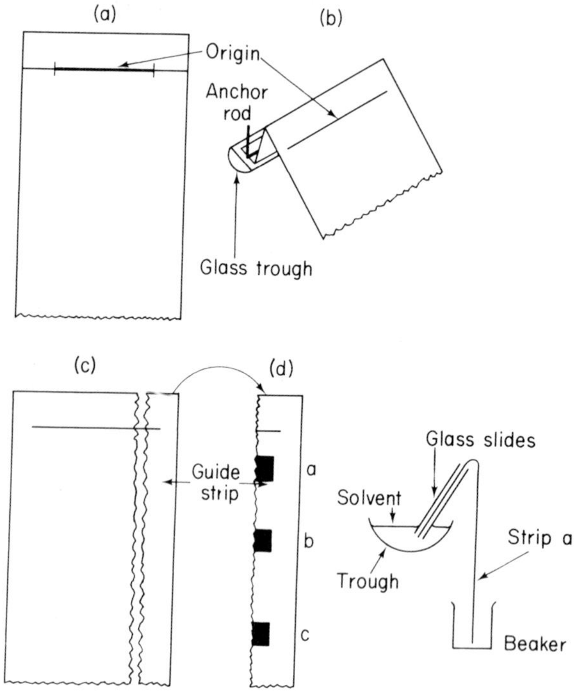

Fig. 8-5. The technique of preparative paper chromatography. (a) A large sheet of chromatography paper (Whatman 3 mm) positioned for sample application. The peptide mixture is streaked along the origin with the use of a capillary pipet keeping the width to about 0.5 cm. A cool airstream dries the sample at the region of application. (b) The chromatography paper is introduced into the chromatography apparatus for descending chromatography. The solvent is placed in the trough and the unit is sealed during the irrigation of the paper. (c) The chromatogram is air-dried and a small section of the sheet is removed as a guide strip to reveal the positions of the separated components with the ninhydrin reagent. (d) The guide strip is matched to the chromatogram and strips (a, b, c) are cut for each zone. (e) The peptides are recovered from the paper strips by descending elution with water or some other appropriate solvent.

PAPER ELECTROPHORESIS

The separation of peptides by the electrophoretic technique, with paper as the supporting medium, depends on differences of net charge that exist among peptides at a given pH. Resolution is high, aided in part by the hydrophobic interaction of the peptides with the paper, and optimum conditions may be sought by trying buffer systems differing in pH.

The apparatus required for paper electrophoresis is extremely simple. A long strip or sheet of filter paper is placed on a glass plate with both ends of the paper immersed in buffer vessels [Fig. 8-6(a)]. Platinum electrodes attached to each vessel are connected to the terminals of a high-voltage power supply. The buffer selected for the separation is used to wet the paper strip. The sample is placed in the center of the paper, the high voltage applied, and the separation is

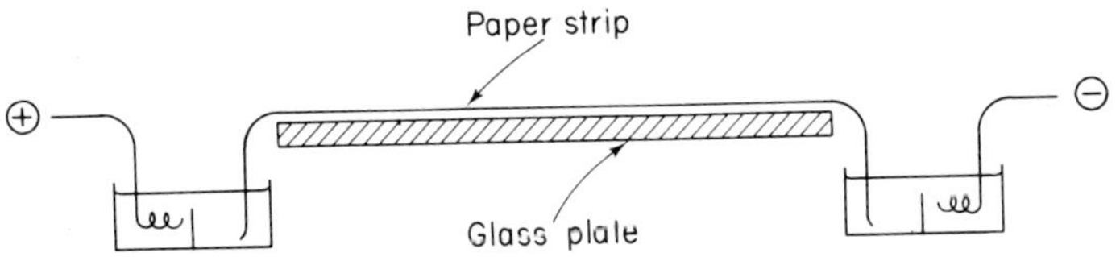

Fig. 8-6(a). The flat-plate electrophoresis apparatus. A sheet or strip of chromatography paper is placed on a glass plate, and the sample is applied along a line perpendicular to the long edge. The paper is wetted with buffer, blotted gently, and the free edges immersed in the buffer vessels. Another glass plate is placed on top of the paper to serve as a pressure plate. The electrodes are attached to a high-voltage power supply.

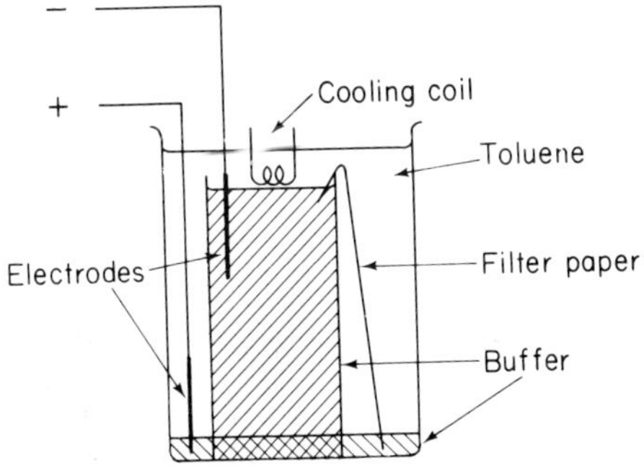

Fig. 8-6(b). High-voltage electrophoresis in a vertical system. A sheet of filter paper is mounted on a plastic frame. The unit is placed in the tank with the upper and lower edges of the paper immersed in buffer and an organic solvent (toluene or other high flash point solvents) covering the remainder of the paper. Platinum electrodes connect to a high-voltage power supply. Cold water circulating through the cooling coil dissipates heat. [Adapted from H. Michl, *Monatsh. Chem.*, **82**, 489 (1951).]

completed in a few hours or less. Heat is generated in the paper and cooling systems are employed to keep the temperature from increasing.

Figure 8-6(b) shows a second type of apparatus in common use. The major difference from the above is the use of inert hydrocarbons as a coolant. The paper sheet is held in the apparatus by a plastic rack, which simplifies the initial setup and the subsequent removal of the sheet. The separated peptides are detected on a guide strip treated with the ninhydrin reagent and their elution from paper follows the procedure mentioned earlier for paper chromatography.

COMMENTS AND RECAPITULATION

The resolution of a peptide mixture is difficult and time-consuming. The relative size, composition, net charge, solubility, and other properties usually vary from one peptide mixture to another, creating a situation that makes it difficult to design a universal approach for the purification of all peptide mixtures. To a large extent, empirical considerations still dominate the strategy of the operations, and the experience of others now provides the rationale for the development of a logical and ordered approach.

Fractionation of samples by column-chromatographic procedures rather than by chromatography or electrophoresis on paper offers several advantages that make it the method of choice, particularly in the early stages of any purification scheme. Depending on the amount of a peptide mixture that must be resolved, the size of the column system may be readily adjusted to operate optimally, since the column dimensions are easily varied. The sample may be introduced onto the chromatographic media with either manual or automatic sample injection systems. Once the separation begins, it is easy to follow the progress of the separation by applying colorimetric procedures or ultraviolet spectrophotometry to monitor the column effluent. If radioactively labeled peptides are present, they are detected with flow-through scintillation counters. Simultaneously, changes in the ionic strength and the pH of the effluent may be followed with in-line conductivity and pH measurements. The time required to complete a separation may be shortened with the use of high flow rates; the flow rates may be obtained with mechanical pumps, which also maintain a constant flow rate and increase the reproducibility of column separations. A wide selection of solvent systems is available, and variations may also be applied to improve the separation by employing a gradient former.

Several chromatographic procedures may be operated in a sequential fashion—in a cascade mode—in which fractions from one procedure are coupled to a second column-chromatographic system or subjected to recycling chromatography. These systems may be controlled automatically by a digital computer programmed to make on-line decisions during a fractionation. Selections are made of fractions that require additional treatment, and decisions can be reached on the mode of further fractionation. Automatic cascade chromatography performs the same operations and separations that are part of the manual approach, but automatic operation decreases the time and manpower requirements.

Gel filtration is the initial step for the fractionation of a peptide mixture. The components are well resolved and recovered in almost theoretical yields. A wide range of solvents offers the possibility of selecting a suitable system to eliminate solubility problems and to avoid unfavorable interactions among the components. It may also be possible to choose a solvent system that is compatible with the system required in the next stage of purification, thereby minimizing time and effort.

Further separation is best accomplished with ion-exchange resins. Dowex 50, operated with a continuous gradient in pH and ionic strength, separates components into a large number of fractions, with the separation based on charge differences and adsorptive effects. Acidic peptides, in particular, and also other fractions that were partially resolved on columns of Dowex 50 resin may be separated on columns of Dowex 1. The sequential use of a cation exchanger followed by an anion exchanger will most probably result in the complete resolution of a peptide mixture. The binding affinity of peptides toward Dowex 1 will clearly be different from their binding affinity toward Dowex 50.

Paper chromatography and paper electrophoresis may be used in the later part of a purification scheme because of the simplicity of the operations, the short time required, and the ability to judge the purity of the separated fractions. It should be noted that paper electrophoresis is also used to estimate the amide content of a peptide. At values greater than pH 5.5, both glutamyl and aspartyl residues are fully ionized having a negative charge while their corresponding amides are electrically neutral. A knowledge of the net charge of a peptide at pH 6.5, for example, in conjunction with the amino acid composition will be sufficient to decide whether the peptide contains a side chain carboxyl or an amide.

In a chapter dealing with the purification of peptides, it is fitting to include a few remarks concerning tests of their purity. It is common to monitor the separation of peptide fractions by paper chromatography and paper electrophoresis. Generally two or more solvent systems for the former and at least two pH values for the latter would be necessary before accepting the presence of a single component as an indication that the sample was homogeneous.

Quantitative amino acid analysis of a peptide is another useful test of purity. A peptide is considered pure if the molecular ratios of amino acids are close to whole numbers. The precision of amino acid analysis is now more than adequate and the sensitivity sufficiently high so that the method serves as a good test of homogeneity.

REFERENCES

Books

C. H. W. Hirs and S. N. Timasheff (Eds.), "Enzyme Structure," *Methods in Enzymology,* Vol. XXV, Part B. Academic Press, New York, 1972.

C. H. W. Hirs (Ed.), "Enzyme Structure," *Methods in Enzymology*, Vol. XI. Academic Press, New York, 1967.

Reviews and Articles

G. M. Edelman and W. E. Gall, "Cascade Chromatography and Automated Multidimensional Fractionation," *Proc. Nat. Acad. Sci.* (*USA*), **68**, 1444 (1971).

C. B. Kasper, "Fragmentation of Proteins for Sequence Studies and Separation of Peptide Mixtures," in *Protein Sequence Determination*, S. B. Needleman (Ed.). Springer-Verlag, New York, 1970.

R. T. Jones, "Automatic Peptide Chromatography," *Methods of Biochem. Analysis,* **18**, 205 (1970).

J. Porath, "Cross-linked Dextrans as Molecular Sieves," *Adv. Prot. Chem.,* **17**, 209 (1962).

Sequence Analysis

NINE

The sequence of a peptide is derived from the combined use of several different degradative techniques, with preference given to those methods that are capable of degrading a peptide in a sequential fashion. Generally, a peptide can be degraded sequentially, starting at either the amino-terminus or at the carboxyl-terminus. Both approaches are commonly tried in an attempt to simplify the sequence analysis and to include as much overlapping data as possible. Although methods of high reliability are used, it is still common practice to apply procedures that duplicate results and that tend to confirm the sequence assignment as the work progresses.

AMINO-TERMINAL METHODS

Sanger's method

In 1945 Sanger developed a successful procedure for the determination of the amino-terminal residue of a protein or peptide. He used the reagent 2,4-dinitrofluorobenzene (DNFB), which reacts with primary amino groups (and also side-chain phenolic, thiol, and imidazole groups) in a Schotten-Bauman type reaction as shown in Eq. (9-1). The dinitrophenyl derivative (DNP-peptide) of the amino-terminal residue is stable to the usual conditions of acid hydrolysis, and the DNP-amino acid is released on complete hydrolysis of the peptide derivative [Eq. (9-2)].

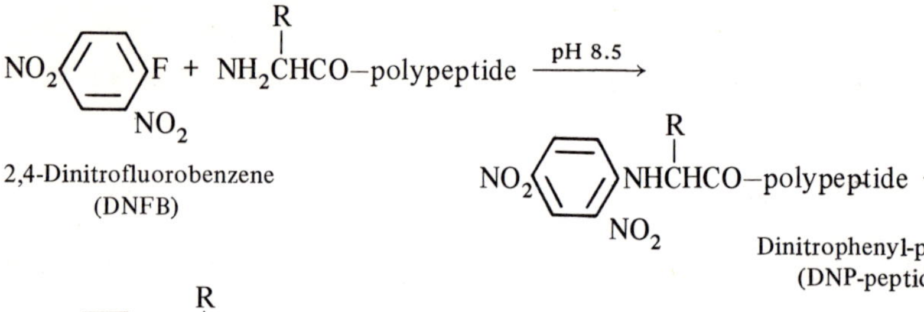

(9-1)

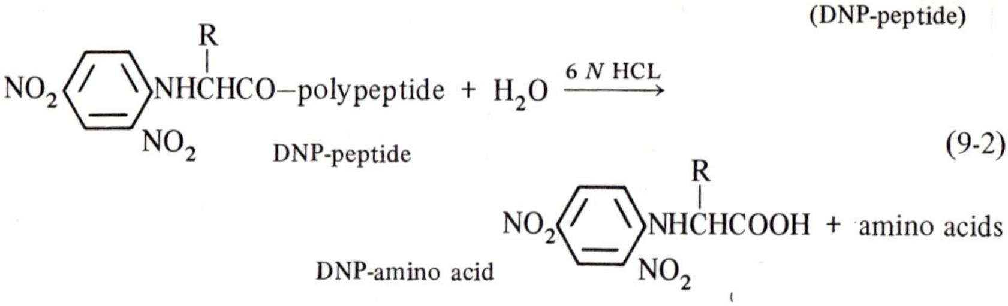

(9-2)

The DNP-amino acid is selectively removed from the remaining amino acids when the hydrolysis mixture is extracted with an organic solvent, such as ether. Identification of the DNP-amino acid is based on its behavior in several paper chromatographic systems with that of known DNP-amino acids. DNP-amino acids are yellow compounds, and their yield is estimated from spectrophotometric analyses at a wavelength of 360 nm of an alkaline solution of the purified sample.

The procedure is limited to the identification of amino-terminal residues of polypeptide chains. The method has been applied to proteins as well as peptides, and useful information may be gained on the subunit structure of a protein (Table 3-3). In his studies on the insulin molecule, Sanger found equal amounts of DNP-glycine and DNP-phenylalanine in hydrolysates of DNP-insulin. This fact indicated that the insulin molecule was composed of two polypeptide chains, a concept that was later proven to be correct.

If the DNP-protein (or DNP-peptide) is submitted to partial acid hydrolysis, a mixture of small DNP-peptides and free amino acids is produced. The composition of DNP-peptides serves to identify those amino acids adjacent to the amino-terminal residue. For example, if one finds DNP-ala, DNP-ala(tyr), and DNP-ala(tyr, ala), the partial sequence of Ala·Tyr·Ala is clearly established. It may be possible to deduce a partial sequence of up to four or five residues from the composition of DNP-peptides present in the hydrolysate.

Cyanate method

In 1963 Stark and Smyth described a new method for the determination of amino-terminal residues, a procedure that is now referred to as the cyanate method. The protein dissolved in a denaturing solvent, such as 8 *M* urea and 4 *M* guanidinium chloride, is treated with potassium cyanate under conditions in which amino groups are converted quantitatively to the carbamyl derivative [Eq. (9-3)].

$$N{=}C{=}O^{\ominus} + NH_2\overset{\overset{R}{|}}{C}HCO\text{–peptide} \xrightarrow{pH\ 8} H_2N\text{–}CO\text{–}NH\overset{\overset{R}{|}}{C}HCO\text{–peptide} \quad (9\text{-}3)$$

Cyanate — Carbamyl peptide

Next, the isolated carbamyl protein is submitted to a cleavage step, a short treatment with strong acid at elevated temperatures, a process that hydrolyzes many peptide bonds and also converts the carbamyl derivative of the amino-terminal residue to the cyclic hydantoin structure [Eq. (9-4)].

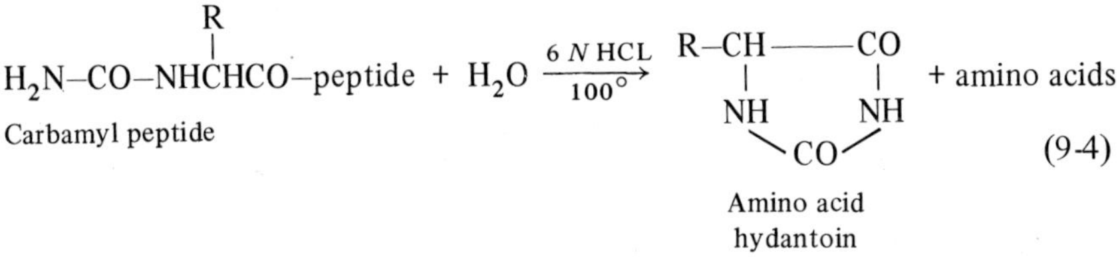

(9-4)

The amino acid hydantoins differ from free amino acids in lacking charged substituents, a difference exploited by ion-exchange chromatography, which readily separates the hydantoin from free amino acids. The isolated hydantoin is hydrolyzed with 6 *N* HCl to regenerate the amino acid [Eq. (9-5)].

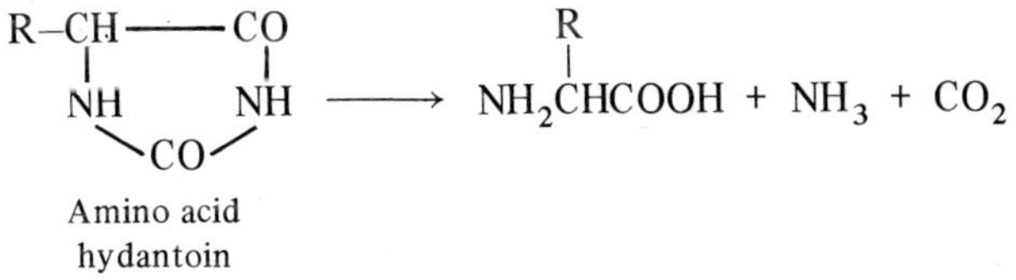

(9-5)

The regenerated amino acid is easily identified on amino acid analysis, and the amount and yield can be readily calculated.

In contrast to the Sanger procedure, the cyanate method depends on the regeneration of the amino acid. The simplicity of determining an amino acid directly, instead of as an amino acid derivative, has contributed to the acceptance and use of the cyanate method with proteins as well as with peptides.

Edman degradation

Edman introduced a method for the sequential degradation of polypeptides that probably represents the single most important technique presently available for sequence analysis. The removal of the terminal residue is accomplished in two steps. In the first, a peptide is treated with phenylisothiocyanate at pH 8, which converts the terminal residue to a phenylthiocarbamyl derivative (PTC-peptide), as shown in Eq. (9-6).

1st Step: Coupling

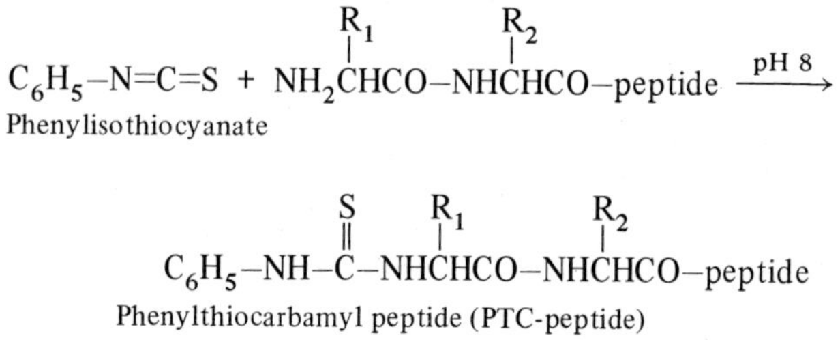

(9-6)

In the second step, the PTC-peptide is treated with anhydrous trifluoroacetic acid, which catalyzes the cyclization of the terminal residue to an anilino-thiazolinone derivative [Eq. (9-7)].

2nd Step: Cyclization

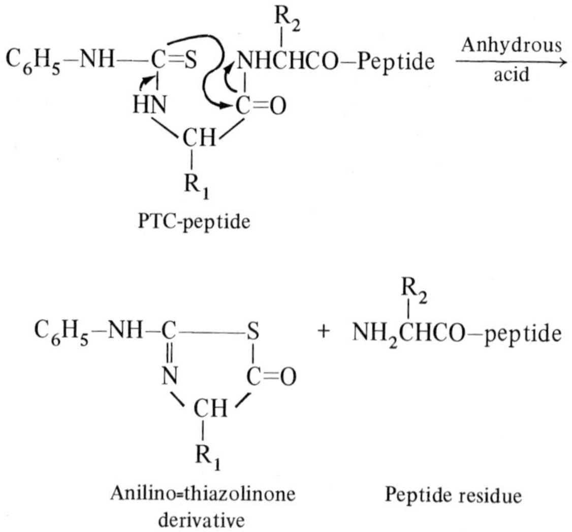

(9-7)

Formation of the five-member ring proceeds by a nucleophilic attack of the thiocarbamyl sulfur atom at the carbonyl of the first peptide bond. Peptide bond cleavage between the first and second residue accompanies ring closure; all remaining peptide bonds stay intact under these nonhydrolytic conditions. The anilino-thiazolinone of the terminal residue is easily separated from the peptide residue on extraction with an organic solvent. The peptide residue is further degraded by repeating the coupling and cyclization steps, continuing the process if necessary, until all residues are removed.

The sequence of amino acids of a polypeptide is established in the Edman degradation when the residue removed at each stage is identified. The amino acid removed is either determined directly as a cyclic derivative or indirectly from the difference in the amino acid composition of the peptide and the peptide residue.

PTH derivative

The direct identification of the amino acid removed is made with the phenyl thiohydantoin derivative (PTH-amino acid), which is formed by an acid-catalyzed rearrangement of the anilino-thiazolinone amino acid [Eq. (9-8)].

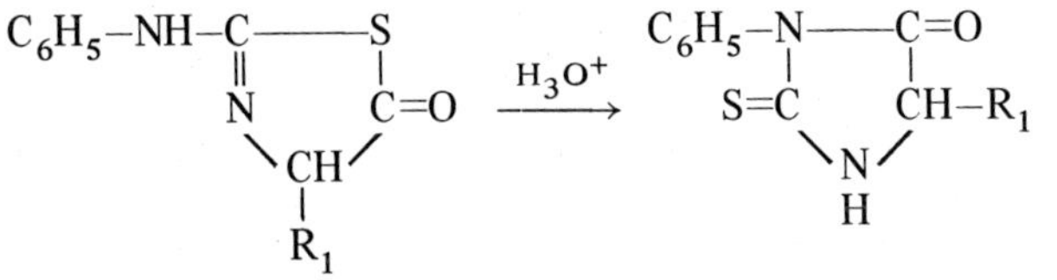

(9-8)

Anilino-thiazolinone amino acid — Phenylthiohydantoin amino acid (PTH-amino acid)

PTH-amino acids are characterized by partition chromatography, with the use of paper or thin-layer chromatography. They are detected on the supporting media by their absorption of ultraviolet radiation and their ability to fluoresce, and the amount present is estimated with the use of the molar extinction coefficient at 265 to 270 nm. Gas chromatography and mass spectrometry are also used for quantitative analysis of PTH-amino acids.

One of the advantages of the direct identification of PTH-amino acids is that it consumes very little sample as a result of the great sensitivity (nanomole range) of the detection systems. A second advantage is seen in the capability of the chromatographic systems to distinguish between glutamine and glutamic acid, as well as asparagine and aspartic acid. The identification of the amides, and other acid-labile amino acids, in the Edman degradation is usually an important factor in assigning these residues in a sequence, since other analyses are generally performed on acid-hydrolyzed samples. The parent amino acid may also be regenerated on treatment of the PTH-derivative with 6 *N* HCl for 16 hours at an elevated temperature of 150°C. The identification of the free amino acid is conveniently accomplished with an amino acid analyzer.

Difference analysis

The amino acids removed at each stage of the degradation may also be identified by amino acid analysis, not of the component removed but of the part remaining. Amino acid analysis of the intact peptide and of the peptide residue after each stage of the degradation is sufficient to establish clearly the sequence of the peptide. The example given in Table 9-1 should make it clear how the sequence of a peptide obtained from human hemoglobin was determined by the use of difference analyses. The molar ratios of amino acids of the intact peptide

TABLE 9-1

Edman Degradation of a Tryptic Peptide of Human Hemoglobin[a]

The composition of each peptide after cyclization is expressed as amino acid residues per mole of peptide. The residue lost at each stage is in **bold face** type.

Stage	*Yield (percent)*	*Val*	*Leu*	*Ser*	*Pro*	*Ala*	*Asp*
Without column[b]							
0		0.95	0.99	0.98	1.04	1.03	0.99
1		**0.10**	0.98	0.99	1.01	1.04	0.98
2		0.05	**0.29**	0.91	1.00	1.00	0.98
3		0.00	0.19	**0.36**	0.99	1.01	0.91
4		0.00	0.15	**0.40**	**0.76**	1.00	0.74
With column[c]							
1	90	**0.05**	1.03	0.92	1.10	1.02	1.02
2	90	0.00	**0.06**	0.91	1.04	1.02	0.97
3	60	0.00	0.02	**0.08**	1.07	1.02	0.94
4	60	0.00	0.00	0.06	**0.10**	1.00	0.91
5	85	0.00	0.00	0.06	0.04	**0.12**	1.00

[a]W. Konigsberg and R. J. Hill, *J. Biol. Chem.*, **237**, 2547 (1962).
[b]The sample was carried through successive stages.
[c]The sample was submitted to purification on a column of the ion-exchange resin Dowex 50-X2, after each stage.

are shown on the first line. Listed below, the molar ratios are given for analyses after the first five stages. The sequence of amino acids is obvious from these results, and the yield of the modified peptide may also be calculated at each stage of the degradation.

One of the advantages of difference analyses is that it employs the amino acid analyzer. Thus supplementary use of suitable procedures for the direct identification of amino acid derivatives is avoided. A disadvantage, however, is that a fraction of the total peptide is consumed at each stage for amino acid analysis, and the amount of peptide remaining may soon become a limiting factor.

Dansyl-Edman procedure

Gray and Hartley developed a highly sensitive amino-terminal end-group method, called the "dansyl" procedure, for monitoring the progress of the Edman degradation. In one variation the peptide sample is distributed in a series of reaction vessels, one for each residue. The Edman degradation is performed sequentially with all samples, but after each stage one sample is removed for end-group analyses. The sequence of the peptide itself is established from the amino-terminal residues of the intact peptide and of the peptide residue following each stage of degradation.

The amino-terminal analyses are performed with the reagent 1-dimethylaminonaphthalene-5-sulfonyl chloride (dansyl chloride, DNS-Cl). The method is similar to Sanger's DNFB procedure, and the reaction of DNS-Cl and peptide produces an acid-stable sulfonamide derivative of the free amino-terminal residue [Eq. (9-9)].

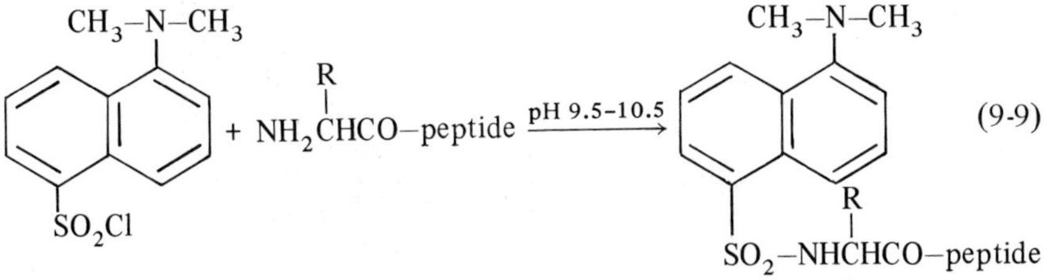

(9-9)

Dansyl chloride   Dansyl-peptide (DNS-peptide)

Hydrolysis of the DNS-peptide with 6 *N* HCl produces free amino acids and the sulfonamide derivative of the amino-terminal residue. The amino acid derivative is then submitted to paper electrophoresis or thin-layer chromatography. Dansyl amino acids (DNS-amino acids) are detected with ultraviolet radiation and are visualized as fluorescent spots. The limits of detection are 10^{-4} to 10^{-5} micromoles, and the sensitivity of the analytical procedure is so great that samples of approximately 5 to 10 nanamoles of peptide are suitable for the Edman procedure (one nanamole is equal to 10^{-9} moles).

Protein-peptide sequenator

In 1967 Edman and Begg described an apparatus called a "protein sequenator," a programmed machine designed to perform the phenylisothiocyanate degradations automatically on samples of proteins and large peptides. (See Fig. 9-1.) The development of an automated procedure depended on each step being performed with a yield of 99 percent or better. Otherwise accumulation of carry-over from previous steps increases the "background noise" and makes a positive identification of a residue questionable.

The critical part of the apparatus is the design and operation of the reaction vessel, which is a spinning glass cup placed in a temperature-controlled glass chamber (Fig. 9-2). The sample is introduced at the base of the cup, and the centrifugal force distributes the sample on the sides as a thin film. The large surface area increases the efficiency of the reactions and the several extractions required and decreases the time required to dry the sample before starting the next stage. Extraction is facilitated by the centrifugal force, which causes the solvent to flow over the surface of the sample to the upper part of the cup, reaching a groove where it is removed by a glass scoop (Fig. 9-2). After each degradative stage, the anilino-thiazolinone derivative is removed by extraction and transferred to a fraction collector for analysis at a later time.

The sequenator has storage units for large reagent bottles containing solutions and solvents required for the degradation. Automatically operated selector values

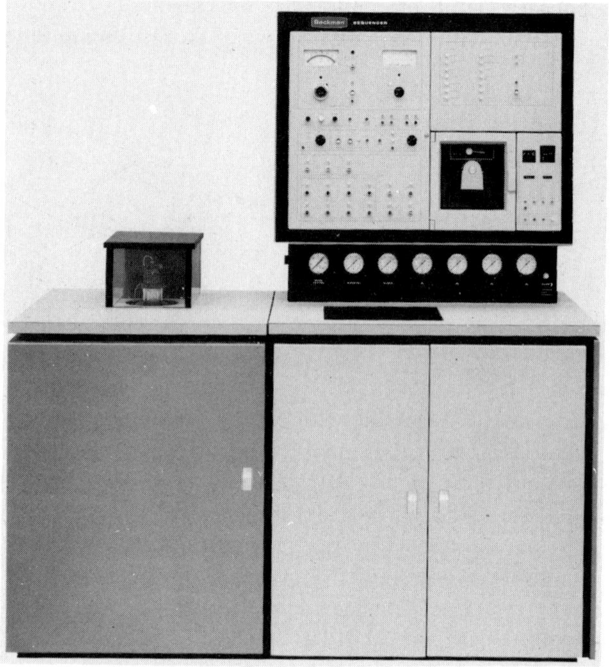

Fig. 9-1. A commercially available protein sequenator. [Illustration courtesy of Beckman Instruments, Inc.]

connect the appropriate liquid system to the reaction vessel (Fig. 9-3). A programmer times all operations and controls the sequence of events for the complete degradation of the sample.

The sequence of the first 60 amino acids of whale myoglobin was reported by Edman and Begg as a demonstration of the capabilities of the instrument. Since the time when they described the sequenator, the unit has been modified, the chemical steps improved, and both protein and peptide samples have been successfully degraded. Although the complete sequence of a protein is not determined directly because of the large size of the molecule, the method appears to be most promising for the sequence determination of up to 50 residues.

Hydrolysis with aminopeptidases

A partial sequence of a peptide may be deduced from a kinetic analysis of amino acids released when a peptide is partially hydrolyzed with either leucine aminopeptidase or aminopeptidase M. These enzymes catalyze the hydrolysis of the peptide bond of the amino-terminal residue, releasing a free amino acid (Chap. 4). Hydrolysis proceeds sequentially from residue to residue, for the degradation continually produces a new amino-terminus [Eq. (9-10)].

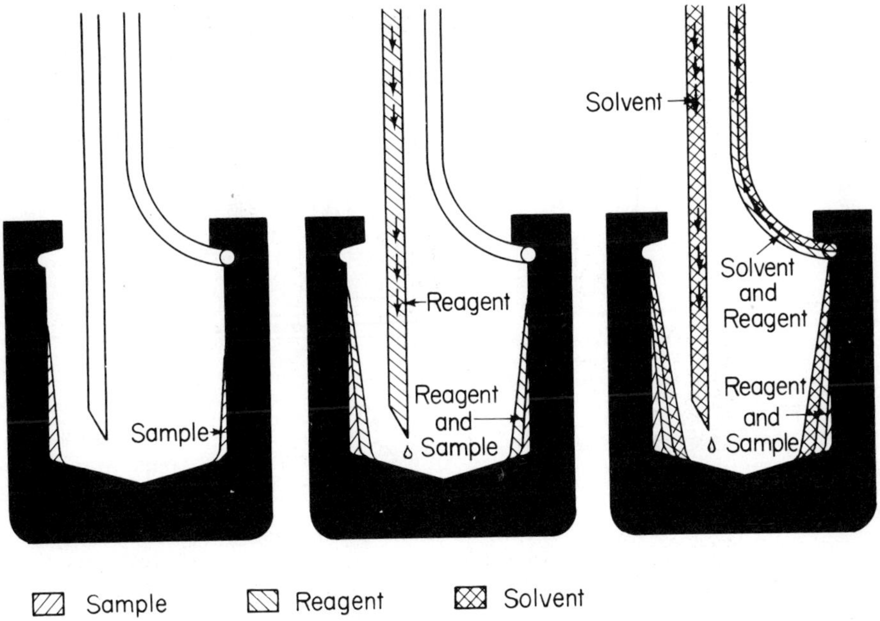

Fig. 9-2. Cross-section of the spinning reaction cup. (a) Protein forms a thin layer in lower half of cup; (b) reagent is added and covers the protein layer; (c) solvent carries liquid up the side of the cup and it is removed by the glass scoop. [Illustration courtesy of Beckman Instruments, Inc.]

$$\mathrm{NH_2\overset{\overset{R_1}{|}}{C}HCO{-}NH\overset{\overset{R_2}{|}}{C}HCO{-}NH\overset{\overset{R_3}{|}}{C}HCO\cdots \xrightarrow[25\text{-}37^{\circ}C]{pH\ 8}} \tag{9-10}$$
$$\mathrm{NH_2\overset{\overset{R_1}{|}}{C}HCOOH + NH_2\overset{\overset{R_2}{|}}{C}HCOOH + NH_2\overset{\overset{R_3}{|}}{C}HCOOH + \cdots}$$

Two forms of the enzyme have been purified from swine kidneys: leucine aminopeptidase is associated with the soluble components of the cell, whereas aminopeptidase M is a particle-bound enzyme present in the microsomal fraction of the cell. Both enzymes are commercially available, suitable for the limited degradation of peptides but slightly different in specificity (Table 9-2). Rapid hydrolysis is observed with aliphatic and aromatic amino acid residues, slower rates with all others, with individual differences spanning a range of several orders of magnitude.

Aminopeptidase M catalyzes the hydrolysis of all peptide bonds, whereas leucine aminopeptidase catalyzes the hydrolysis of all but the imide bond of proline. Consequently, hydrolysis of a peptide with leucine aminopeptidase removes amino acids that are amino-terminal to the imide bond, and, thus, the

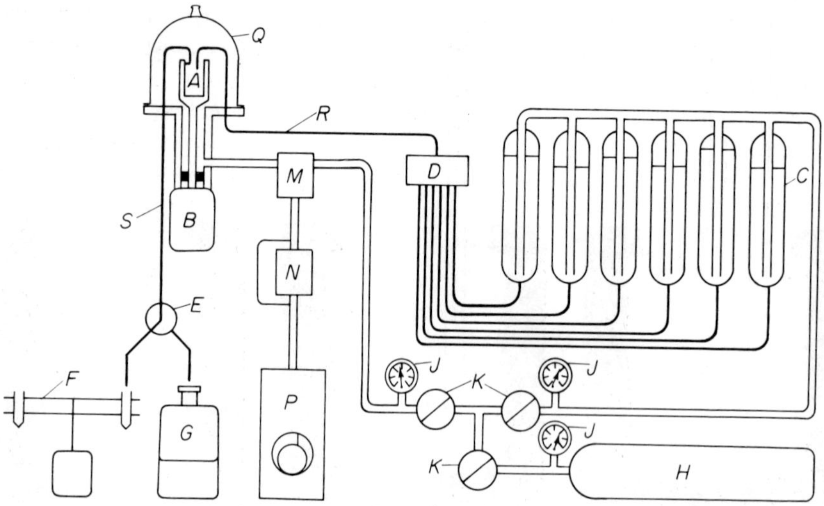

Fig. 9-3. A schematic diagram of the protein sequenator. (a) Spinning cup; (b) electric motor; (c) reagent (solvent) reservoir; (d) value assembly; (e) outlet stopcock assembly; (f) fraction collector; (g) waste container; (h) nitrogen cylinder; (j) pressure gauges; (k) pressure regulators; (m) 3-way valve; (n) 2-way valve with bypass; (p) rotary vacuum pump; (q) bell jar; (r) feed line; (s) effluent line. Gas lines are doubly contoured, and liquid lines are filled. [From P. Edman and G. Begg, *European J. Biochem.*, **1**, 80 (1967).]

TABLE 9-2

Rates of Hydrolysis of Amino Acid Derivatives by Aminopeptidases

	Leucine Aminopeptidase[a,b]	*Aminopeptidase M*[c,d]	
	Amide	*Amide*	p-*Nitroanilide*
Glycine	0.33	0.085	68
Alanine	3.4	1.15	100
Valine	17	0.11	
Isoleucine	20		
Leucine	100	0.32	71
Tryptophan	24		
Phenylalanine	26	0.29	83
Tyrosine	16		
Histidine	19	0.058	
Lysine	7.1		
Arginine	7.1		
Serine	0.76	0.2	
Proline	0.71	0.004	

[a]E. L. Smith and D. H. Spackman, *J. Biol. Chem.*, **212**, 271 (1955).

[b]The enzyme was activated with Mn^{2+}. The value for leucinamide is given as 100 and for the other amides as the relative rate.

[c]E. D. Wachsmuth, I. Fritze, and G. Pfleiderer, *Biochemistry*, **5**, 175 (1966).

[d]Values are expressed as percentage of V_{max} of alanine-*p*-nitroanilide.

resulting peptide residue begins with the sequence X-Pro, and "X" may be one of the normally occurring amino acids. The enzyme prolidase is capable of hydrolyzing the X-Pro linkage, and it must be added along with leucine aminopeptidase if the releasing of the remaining amino acids is desired.

The limited degradation of a peptide with an aminopeptidase releases amino acids in fractional amounts, and the example given in Table 9-3 shows how a partial sequence is established. In addition to providing a partial sequence, or an

TABLE 9-3

Time-dependent Release of Amino Acids from a Peptide Submitted to Hydrolysis with Leucine Aminopeptidase[a]

(The amino acids released are expressed as amino acid residues.)

Time (minutes)	*Ala*	*Asn*	*Gly*	*Val*	*Glu*	*Trp*
3	0.95	0.60	0.20	0.08		
10	1.40	1.00	0.62	0.40		
30	1.95	1.05	1.02	0.91	0.75	

Sequence of peptide Ala-Asn-Gly-Val-Ala-Glu-Trp

[a]K. Weber and W. Konigsberg, *J. Biol. Chem.*, **242**, 3563 (1967).

ordering of residues, the release of amino acids also gives valuable information on the location of the acid-labile residues of asparagine, glutamine, and tryptophan. Since these residues are released intact, their direct identification is more reliable than the indirect assignment made from electrophoretic mobilities on the intact peptide (amide or acid group) or from spectrophotometric analyses for tryptophan.

CARBOXYL-TERMINAL METHODS

Hydrazinolysis method

The chemical procedure most commonly used for the identification of the carboxyl-terminal residue of a protein or peptide is hydrazinolysis. The sample is dissolved in anhydrous hydrazine and the mixture is kept at 100 degrees for 20 hours. The carboxyl-terminal residue is released as the free amino acid; all other residues are present as the strongly basic amino acid hydrazides [Eq. (9-11)]. The carboxyl-terminal amino acid may be separated from the hydrazides by the use of Sanger's DNP-procedure or by ion-exchange chromatography. The terminal residue may be identified with the use of an amino acid analyzer, but the results must be considered only qualitative, for recoveries are usually poor. Unfortunately, the yield varies with the protein under investigation, and optimum conditions for the reactions are difficult to obtain. Furthermore, amino acid

$$\underset{\text{Polypeptide}}{NH_2\overset{R}{\overset{|}{C}}HCO{-}(NH\overset{R}{\overset{|}{C}}HCO)_n{-}NH\overset{R}{\overset{|}{C}}HCOOH} + \underset{\text{Hydrazine}}{NH_2NH_2} \longrightarrow$$

$$(n+1)\ \underset{\text{Amino acid hydrazides}}{NH_2\overset{R}{\overset{|}{C}}HCO{-}NHNH_2} + \underset{\text{Carboxyl-terminal amino acid}}{NH_2\overset{R}{\overset{|}{C}}HCOOH} \qquad (9\text{-}11)$$

hydrazides undergo slow hydrolytic decomposition that increases the content of free amino acids.

Carboxyl-terminal residues with amide groups in their side chains (asparagine and glutamine) or at the α-carboxyl position will also form hydrazides. If this reaction occurs, a free amino acid is not produced on hydrazinolysis; instead all residues appear as hydrazides. If such is the case, the hydrazinolysis method fails unless these hydrazides are isolated and identified. Hydrazine slowly reacts with the guanidinium group of arginine, partly transforming the amino acid to ornithine and increasing the difficulty in establishing arginine as a carboxyl-terminal residue.

Success with the hydrazinolysis method requires quantitative conversion of the amino acid residues to hydrazides, and even a small degree of hydrolysis must be avoided. A large excess of hydrazine helps to drive the reaction to completion, and anhydrous conditions may eliminate hydrolysis.

Hydrolysis with carboxypeptidases

The carboxyl-terminal residue is selectively released from a protein or peptide when the sample is treated with pancreatic carboxypeptidase. The release of amino acids from the carboxyl-terminus complements the sequence information obtained from the use of leucine aminopeptidase, which releases amino acids from the amino-terminus.

Two types of carboxypeptidase have been characterized from the bovine pancreas: the first to be discovered is now referred to as carboxypeptidase A and the second is carboxypeptidase B. Carboxypeptidase A catalyzes the hydrolysis of peptide bonds derived from a large number of different carboxyl-terminal amino acid residues. However, hydrolysis is most rapid with aromatic amino acid residues; other residues are also hydrolyzed, but large differences are observed in their relative rates (Table 9-4). Carboxypeptidase A does not release prolyl residues nor those of lysine and arginine. When susceptible residues are located at the carboxyl-terminus and at neighboring positions, the enzyme releases amino acids sequentially [Eq. (9-12)]. The relative yield of each amino acid must be followed with time in order to deduce the correct sequence.

TABLE 9-4

Rates of Hydrolysis of Peptide Substrates by Carboxypeptidase A[a]

Rapid release	Tyr, Phe, Trp, Leu, Ile, Met, Thr, Gln, His, Ala, Val
Slow release	Asn, Ser, Lys[b]
Very slow release[c]	Gly, Asp, Glu, $CySO_3H$, *S*-carboxymethyl Cys
Not released[c]	Pro, Arg

[a]Adapted from R. P. Ambler, *Methods in Enzymology*, **25**, 262 (1972).

[b]The rate of release of lysine may be influenced by the pH of the hydrolysis mixture.

[c]The presence of a "very slow" or "not released" amino acid as penultimate residue will generally decrease the rate of release of the C-terminal amino acid.

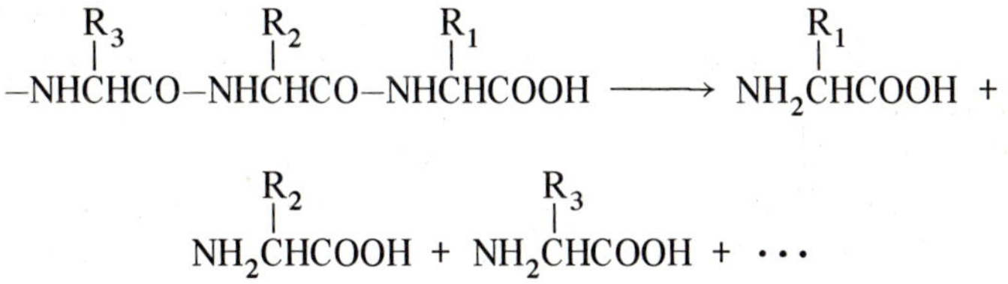

(9-12)

The specificity of carboxypeptidase B is restricted to the hydrolysis of terminal residues of lysine and arginine. If either of these residues is at the carboxyl-terminus, the adjacent peptide bond is hydrolyzed, thereby releasing the basic amino acid [Eq. (9-13)].

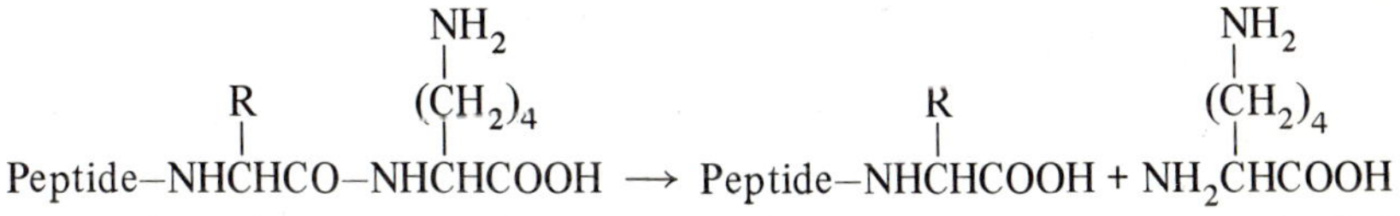

(9-13)

The new carboxyl-terminal residue may then be released if carboxypeptidase A is also present. Both enzymes must be used to release amino acids sequentially from tryptic peptides, for these peptides terminate in a basic residue.

COMMENTS AND RECAPITULATION

In theory, the complete sequence of a peptide could be completed with the judicious application of the Edman procedure for the requisite number of stages. Indeed, a number of examples adequately support this contention and demonstrate the great utility of the method. Unfortunately, many examples are also available where the degradation failed after several residues were removed and the sequence analysis was incomplete.

A successful degradation of a peptide with the Edman procedure requires that two experimental conditions be met; otherwise ambiguities will be experienced when the results are evaluated. The first of these conditions is that the yield of the product must be essentially quantitative at every step of the degradation, and the second is that losses of the peptide due to solubility in extracting solvents and handling be minimal.

The yield may vary from one step to the next, being dependent on the composition of the peptide. For example, the degradation is influenced by steric factors, by electrostatic interactions, and by the stability of the amino acid derivative and of the peptide itself. Ideally, the reaction conditions should be varied at each stage in an attempt to arrive at optimal conditions, but usually this approach is impractical because insufficient sample quickly becomes a limiting factor. If the degradation is incomplete at any stage, a family of peptide residues differing in size will accumulate. If this happens, the mixture of partly degraded peptides may be submitted to a purification scheme; further degradation is then continued with the purified peptide residue.

When a peptide has been degraded through a number of stages, the amount of sample remaining may become very small because of losses due to partial solubility in the solvents used for extraction. The extent of the loss is related to the composition and sequence of the peptide, and it may be appreciable if the peptide has a number of nonpolar residues. It may be possible to minimize losses with a better choice of solvents or with modifications of the procedure that eliminate extraction requirements.

Since the Edman degradation may become unreliable as the number of stages increases, it is common practice to employ other methods that serve to degrade the peptide in a sequential fashion. Carboxypeptidases release amino acids from the carboxyl-terminus that complement the sequence information derived from the Edman degradation. If residues can be placed in a known sequence from their rate of release, a partial sequence can decrease the number of stages required for the Edman degradation. The combined results of an Edman degradation and carboxypeptidase hydrolysis will generally suffice to establish the complete sequence of most peptides.

It may also be necessary to submit large peptides to further degradation in order to produce a mixture of peptides and to use the fragments for further sequence analysis. A peptide that was initially obtained from the tryptic hydrolysis of a protein may be susceptible to the action of chymotrypsin, while peptides from the chymotryptic digest of the protein may be degraded with trypsin (Chap. 7). Furthermore, enzymes that show a broad specificity, such as papain, pepsin, and subtilisin, degrade large peptides to much smaller fragments.

REFERENCES

Books

C. H. W. Hirs and S. N. Timasheff (Eds.), "Enzyme Structure," *Methods in Enzymology*, Vol. XXV, Part B. Academic Press, New York, 1972.

S. Blackburn, *Protein Sequence Determination. Methods and Techniques.* Marcel Dekker, Inc., New York, 1970.

Reviews and Articles

M. A. Hermodson, L. H. Ericsson, K. Titani, H. Neurath, and K. A. Walsh, "Application of Sequenator Analyses to the Study of Proteins," *Biochem.*, **11**, 4493 (1972).

P. Jollès and J. Jollès, "Primary Sequences of Proteins and their Evolution," *Progress in Biophysics and Molecular Biology*, **22**, 127 (1971).

K. Narita, "End-Group Determination," in *Protein Sequence Determination*, S. B. Needleman (Ed.). Springer-Verlag, New York, 1970.

P. Edman, "Sequence Determination," in *Protein Sequence Determination*, S. B. Needleman (Ed.). Springer-Verlag, New York, 1970.

G. R. Stark, "Recent Developments in Chemical Modification and Sequential Degradation of Proteins," *Adv. Protein Chemistry*, **24**, 261 (1970).

W. R. Gray and J. F. Smith, "Rapid Sequence Analysis of Small Peptides," *Anal. Biochem.*, **33**, 36 (1970).

N. Seiler, "Use of the Dansyl Reaction in Biochemical Analysis," *Methods in Biochem. Analysis*, **18**, 259 (1970).

D. Beale, "Amino Acid Phenylthiohydantoins, Dimethylaminonaphthalene Sulphonyl (DANSYL) Amino Acids and Sequence Analysis of Proteins," in *Chromatographic and Electrophoretic Techniques*, I. Smith (Ed.), **1**, 189 (1969).

P. Edman and G. Begg, "A Protein Sequenator," *European J. Biochem.*, **1**, 80 (1967).

Disulfide Bond Assignments and the Diagonal Electrophoretic Technique

TEN

The first step in the sequence analysis of a protein containing disulfide bonds is the selective cleavage of these bonds and the conversion of the sulfur atoms to stable derivatives. Further operations are then performed with the modified protein. As a result, the final amino acid sequence is that of the modified protein, and it is then necessary to establish the correct pairing between appropriate half-cystine residues.

DIRECT ISOLATION OF DISULFIDE-CONTAINING PEPTIDES

The approach that Sanger first employed in his studies with the insulin molecule is still followed by many investigators (Fig. 10-1). In the first step, the protein with intact disulfides is treated with either 8 *M* urea or 6 *M* guanidinium chloride in order to denature the molecule and to increase its susceptibility to subsequent degradation with proteolytic enzymes. Hydrolysis of the denatured protein with proteolytic enzymes and the isolation of disulfide-containing peptides must be performed under a set of experimental conditions that avoid disulfide interchange (Chap. 6). The possible rearrangement of disulfide bridges is kept to a minimum if mercaptans are absent from the system, for these agents act as catalysts in the interchange reaction. Therefore proteins that contain free sulfhydryl groups must be treated with reagents (iodoacetate) that convert the mercaptan to stable derivatives. Furthermore, reaction conditions employed in proteolytic enzyme digestions should avoid high pH values that may cause the cleavage of the disulfide bond; typically, a pH no higher than 7 is used.

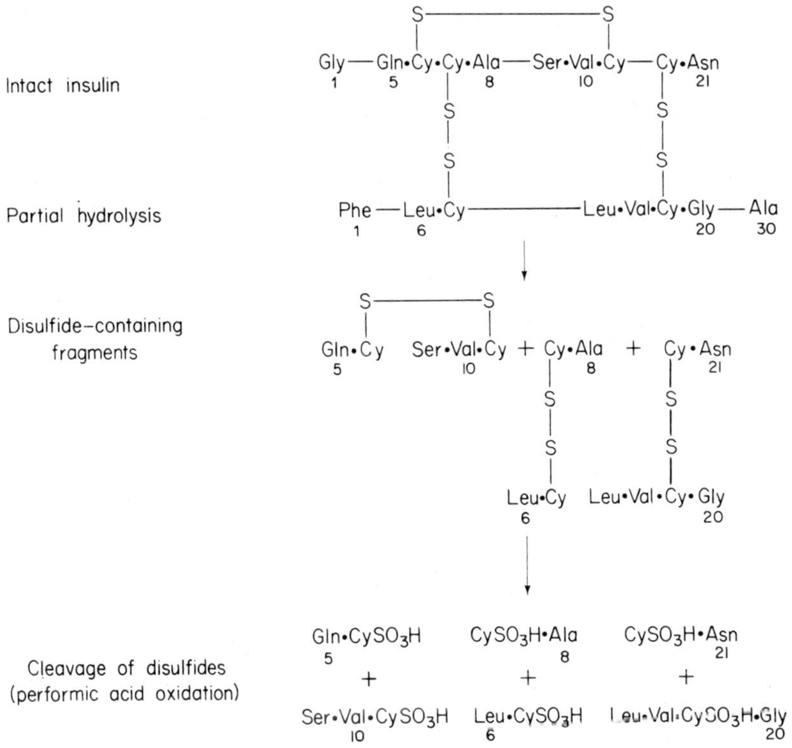

Fig. 10-1. Disulfide bridges in the insulin molecule. The protein was degraded to a mixture of peptides (partial hydrolysis or preferably by extensive proteolysis). The cystine-containing peptides were isolated, and each one was submitted to performic acid oxidation. The structures of the fragments were established and were sufficient to reveal the sequences surrounding the disulfide bonds as they existed in the intact insulin molecule.

The proteolytic enzyme digest contains a mixture of peptides, a few of which are linked to one another by disulfide bonds. The peptide mixture is resolved by techniques discussed earlier (Chap. 8), and specific analytical methods are used to detect peptides with intact disulfide bonds.

Purified disulfide-containing peptides are next cleaved by performic-acid oxidation or by reduction and alkylation. Let us assume that a peptide, on performic-acid oxidation, produces two oxidized peptides. It is clear that these peptides represent the two members of the cystine linkage. All that would be necessary to identify the two peptides is to perform an amino acid analysis on each and to relate the peptide compositions to the amino acid sequence of the protein. In this way the assignments of the disulfide bonds, as they occur in the native protein, are clearly established (Fig. 10-1).

When disulfide bridges were assigned in the sequences of ribonuclease, lysozyme, papain, and in a few other cases, hydrolysis of the denatured protein was performed with pepsin rather than with trypsin and chymotrypsin. Pepsin displays a broad specificity, causing the hydrolysis of many peptide bonds, and the cystine-containing peptides are small sized, thus making their isolation and characterization relatively easy. Furthermore, the use of pepsin, which is catalytically active at acid pH values, helps to minimize potential problems arising from disulfide interchange.

DIAGONAL TECHNIQUE

The disulfide bridges can be assigned in a more direct fashion and with a minimum of work if the diagonal electrophoretic technique developed by Brown and Hartley is substituted for the earlier procedures. The technique is an application of a unique and ingenious way to resolve a peptide mixture, and the same approach may also be applied to the solution of several other problems associated with sequence studies.

The assignment of disulfide bridges by the diagonal technique makes use of three simple operations (Fig. 10-2). In the first, a mixture of peptides is submitted to an electrophoretic separation on paper at a given pH. Next, the paper sheet is treated with vapors of a reagent capable of reacting specifically with the disulfide bond. For example, performic-acid vapors may be used to cleave the disulfide bond with the formation of cysteic acid residues and a corresponding change in the net charge of the oxidized peptides. Then the paper sheet is again submitted to an electrophoretic separation with identical conditions to those used before except that the direction of electrophoresis is now at right angles to the direction used earlier. The chemically modified peptides will have an altered mobility in the second electrophoretic separation, whereas all peptides maintain their original mobility. Consequently, peptides fall on a diagonal line with the exception of the modified peptides, which are displaced from the diagonal. Inspection of the peptide positions of those that are displaced gives directly both the number and the net charge of the altered peptides.

The application of the diagonal technique to the elucidation of disulfide linkages will be described in sufficient detail to provide an appreciation of the events and to lay the groundwork for its use with other specific sequence assignments. In the first step, a denatured protein with intact disulfide bonds is submitted to hydrolysis with pepsin. The resulting peptide mixture is applied to the center of a large sheet of chromatography paper, and the sample is submitted to electrophoresis at pH 6.5 (Fig. 10-2).

The zone containing the peptides is then removed from the sheet by cutting out the appropriate-size strip. The paper strip is placed in a desiccator over a dish containing performic acid. The vapors of performic acid penetrate the strip,

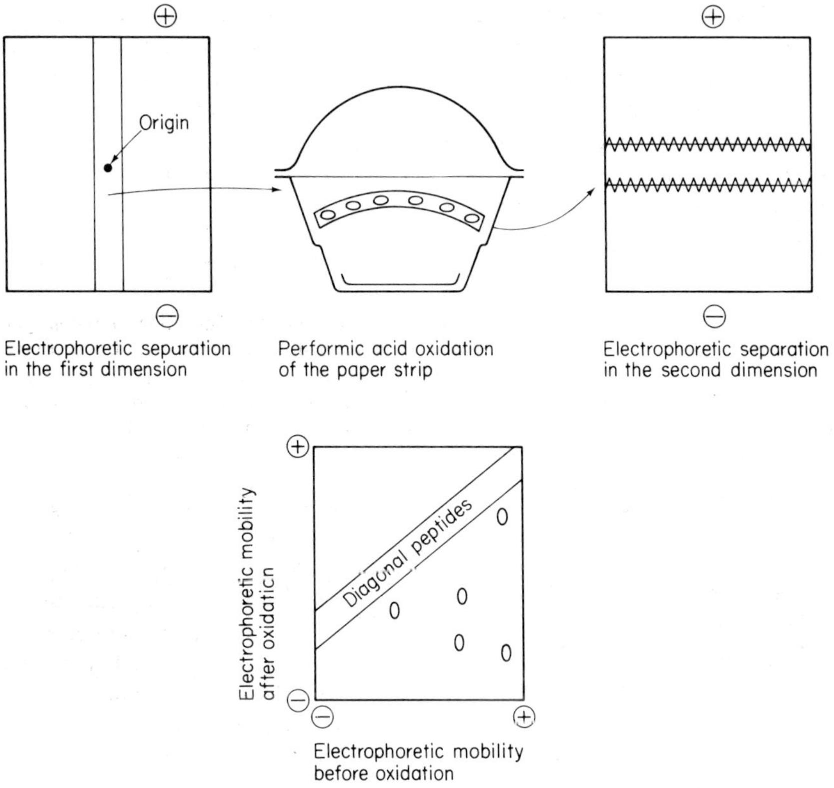

Fig. 10-2. The technique of diagonal electrophoresis.

and in a matter of a few hours disulfides are oxidized to produce the cysteic acid derivatives. Disulfides that bridge two peptides are converted to two new peptides, each now containing an acidic residue [Fig. 10-3(a)]. Disulfides present in a single peptide are oxidized, and a single peptide with an altered charge is formed [Fig. 10-3(b)].

The paper strip is next submitted to a second separation with electrophoresis at pH 6.5 (Fig. 10-2). To accomplish this step, the strip is attached to two large sheets of paper by sewing the two together, which essentially places the peptide mixture once again in the center section of a large sheet of chromatography paper. The electrophoretic separation is then run in a direction perpendicular to the one used in the first separation. Peptides are revealed on dipping the paper sheet into a solution of ninhydrin and allowing the color to develop at room temperature. Peptides are located on the diagonal except for those that initially contained disulfides. These oxidized peptides now have a different mobility when run in the second dimension and appear as spots away from the diagonal (Fig. 10-4).

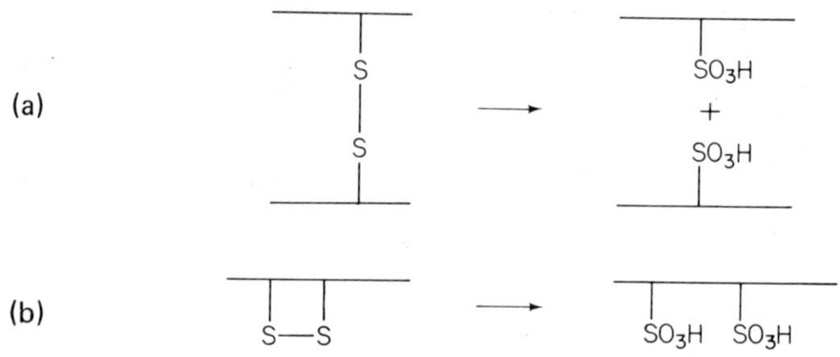

Fig. 10-3. The performic acid oxidation of disulfide-containing peptides: (a) show the conversion of an intermolecular bridge to two peptides; (b) shows the conversion of an intramolecular bridge to a modified peptide. The disulfide bond is oxidized to a sulfonic acid derivative producing two residues of cysteic acid.

The diagonal electrophoretic technique serves as a fast way to gain a qualitative picture of the number and type of disulfide bridges present in the protein molecule. With a small variation of the procedure, the oxidized peptides can be isolated on a preparative scale in a yield suitable for sequence analyses.

The diagonal electrophoretic technique is an ingenious application of simple peptide chemistry (performic-acid oxidation) coupled with a method for the rapid separation of peptides (paper electrophoresis). The simplicity of the operations, the use of very little sample, and the rapidity of the method have contributed to make the procedure an ideal approach for the direct analysis of disulfide-containing peptides.

Other applications

The principle underlying the diagonal technique may be applied to the solution of other problems in sequence analysis. One additional example will illustrate the flexibility of the procedure. It will be recalled that maleic anhydride can be used for the reversible blocking of amino groups [Chap. 7, Eq. (7-1)]. In a tryptic digest of a maleylated protein, peptides contain arginine at the carboxyl-terminal position, in agreement with the specificity of trypsin, which is redirected toward the arginine residues only (Fig. 7-3). The ϵ-maleyl lysyl residues cannot be hydrolyzed by trypsin as long as the ϵ-amino group is blocked.

An electrophoretic separation of the tryptic digest of the malelyated protein is performed at pH 6.5. A strip of paper containing the separated peptides is then cut out and placed in a desiccator over vapors of a pyridine-acetate buffer, at pH 3.5 and a temperature of 60 degrees. Under these conditions, the maleylated derivatives are unblocked in 6 hours, regenerating free amino groups [Fig. 7-4(a)].

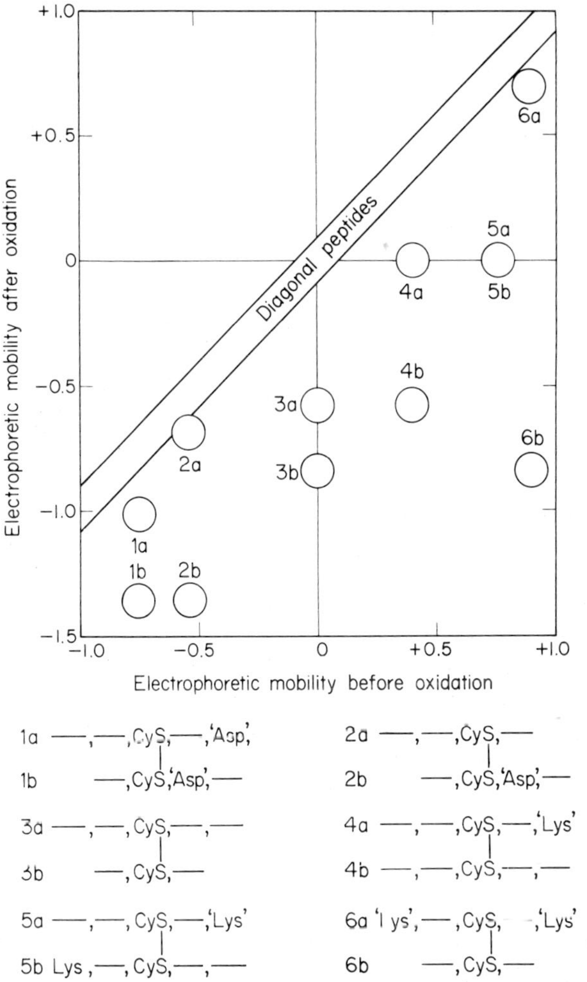

Fig. 10-4. Hypothetical diagonal map produced by some common types of cystine peptides. The blank lines in the sequence represent single amino acids that do not bear charged groups. [From J. R. Brown and B. S. Hartley, *Biochem. J.*, **101**, 214 (1966) Fig 7.]

The strip of paper is sewn onto a fresh sheet of paper, and electrophoresis is carried out at pH 6.5 as before, but at right angles to the original direction.

The removal of the maleyl group results in a change in the charge of the peptide from zero to plus 2 for lysine-containing peptides and from minus one to plus one for the amino-terminal peptide (Fig. 10-5). If two lysyl residues are present in a tryptic peptide, the charge changes from minus one to plus 3. The change in charge between the first dimension and the second displaces peptides

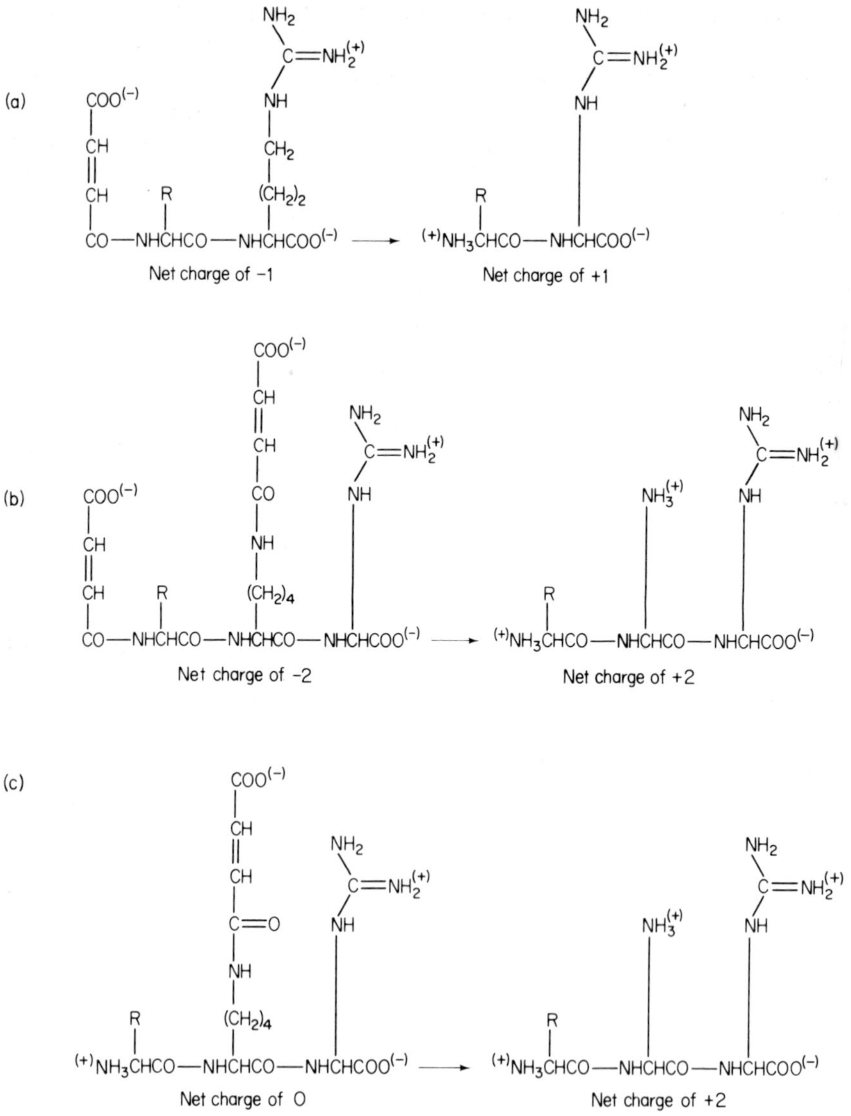

Fig. 10-5. The net charge of maleyl peptides and of deblocked peptides at pH 6.5. (a) Amino-terminal peptide; lysine absent; (b) amino-terminal peptide; lysine present; (c) tryptic peptide.

from the diagonal, and the number of displaced peptides is related to the number of lysyl residues.

The method is easily adapted to operate on a preparative scale, and the unblocked peptide can be recovered after the second electrophoretic separation.

If the peptide is now submitted to hydrolysis with trypsin, cleavage occurs at the lysyl residues. The fragments are separated and the sequence of each is established. It is now relatively easy to place the pieces in their proper order, since the peptide containing arginine must be derived from the carboxyl-terminal position of the intact peptide (Fig. 7-3). The direct alignment of the fragments into their proper order eliminates the use of a second method of degradation and simplifies the operations required to complete the sequence assignment.

REFERENCES

Reviews and Articles

H. A. McKenzie and D. C. Shaw, "Alternate Positions for the Sulphydryl Group in β-Lactoglobulin: The Significance for Sulphydryl Location in Other Proteins," *Nature New Biology,* **238**, 147 (1972).

P. J. G. Butler, J. I. Harris, B. S. Hartley, and R. Lieberman, "The Use of Maleic Anhydride for the Reversible Blocking of Amino Groups in Polypeptide Chains," *Biochem. J.,* **112**, 679 (1969).

L. B. Smillie and B. S. Hartley, "Histidine Sequences in the Active Centres of some 'Serine' Proteinases," *Biochem. J.,* **101**, 232 (1966).

J. R. Brown and B. S. Hartley, "Location of Disulphide Bridges by Diagonal Paper Electrophoresis," *Biochem. J.,* **101**, 214 (1966).

M. A. Raftery and R. D. Cole, "On the Aminoethylation of Proteins," *J. Biol. Chem.,* **241**, 3457 (1966).

L. Lumper and H. Zahn, "Chemistry and Biochemistry of Disulfide Exchange," *Advances in Enzymology,* **27**, 199 (1965).

D. H. Spackman, W. H. Stein, and S. Moore, "The Disulfide Bonds of Ribonuclease," *J. Biol. Chem.,* **235**, 648 (1960).

Structure and Function and the Molecular Evolution of Proteins

ELEVEN

In this chapter we examine the relationships between the amino acid sequence of a protein, its three-dimensional structure, and its biological activity. The elucidation of structure-to-function relationships is clearly the long-term aim of many studies in the field of protein chemistry. Indeed, major contributions have already been made in a number of diverse areas: kinetic measurements of enzyme action; physical studies of the behavior of protein molecules in solution; chemical modifications of proteins; amino-acid-sequence analysis of a protein as it varies by mutation in a given species and as it varies from one biological species to another; amino-acid-sequence analysis of "families" of related proteins and x-ray crystallographic studies of enzymes (and proteins generally) and enzyme-inhibitor complexes. Instead of attempting a broad coverage of the subject, the current status of the problem will be illustrated with a few specific and pertinent details, limiting our examples to a few cases.

The biological activity of a protein is associated with a specific three-dimensional structure of the molecule, which, in turn, is specified by the amino acid sequence (Chap. 1). Our present view of the structure of globular proteins is that most amino acid residues are stabilizing elements and that only a small number participate directly as active elements in the biological activity. In the folding of globular proteins most amino acid residues are positioned below the surface as "buried" residues. It is now believed that buried residues, which are predominantly hydrophobic, provide the stabilization that is required to maintain the three-dimensional structure (Chap. 1). Those residues on the surface are mainly amino acids with polar and ionic side chains. For the most part, these side chains are capable of interacting with surrounding solvent molecules and may bind small molecules or proteins, and thus participate in those events responsible for the biological activity. Recent research has been focused on elucidating the nature, properties, and action of those residues directly concerned with function. We now turn our attention to this phase of the problem.

MECHANISM OF ENZYME ACTION

The active center of an enzyme

From an experimental point of view, one can study the structure-to-function relationships of a protein molecule by chemical modification of surface residues. This point can be made clear from studies carried out with enzymes. In every enzyme-catalyzed reaction, the substrate (whether a large or small molecule) binds with a specific site on the enzyme, forming an enzyme-substrate complex (ES complex). As an ES complex, the substrate molecule may undergo a change in conformation leading to a transition state in which bond breaking and bond making occur, with a transformation of the substrate to products. Since many substrates are small molecules, it is clear that the specific binding to the enzyme (ES complex) may involve only a small region of the enzyme structure (Fig. 11-1). That region of the enzyme that takes part in the interaction with the substrate is called the *active center*. The *active site* is that part of the active center that is concerned with the catalytic event. The *specificity site* is the region of the enzyme, close to the active site, that binds the substrate and properly orients the molecule "to fit" at the active site. These concepts will serve as the basis for a further discussion of enzyme action.

Chemical modification of the active site

The identification of the amino acid residues of the active site makes use of chemical reagents that react with specific groups of the enzyme and that produce a loss of enzyme activity. In order to be certain that inactivation involves a residue or residues of the active site rather than critical structural elements remote from the active site, inactivation by a chemical reagent must be inhibited in the presence of the substrate or a substrate analog. Since a substrate or a competitive inhibitor binds at the active-site region, steric hindrance would effectively block access of the chemical reagent.

Chemical modification of the active site "labels" the critical residues in the form of a chemical derivative. Amino acid analysis of the inactive enzyme identifies directly those amino acid residues that participated in the reaction, and sequence analysis of the enzyme derivative establishes the location of the specific residues that now appear as chemical derivatives.

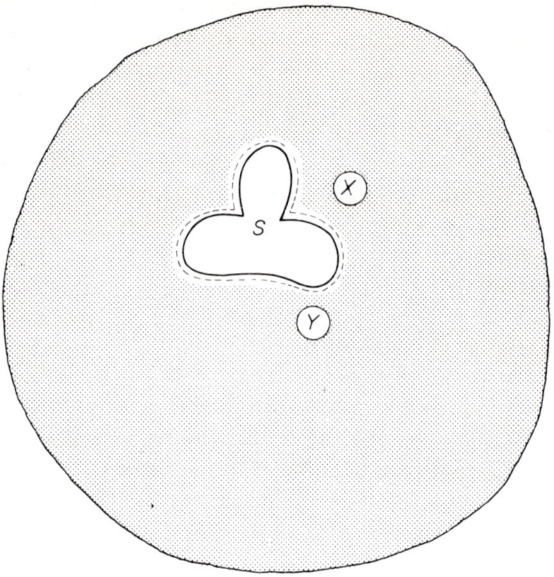

Fig. 11-1. A diagramatic representation of a substrate molecule (labeled S) bound to a specific region of an enzyme. The substrate interacts with the active center, consisting of a distinct topographical area complementary to the structure of the substrate (called the specificity site) and amino acid side chain groups (X and Y) that function as catalysts (called the active site).

We can pursue this topic further with a specific example, and chymotrypsin is an excellent choice from a historical point of view and because of the wealth of information available on its properties and functions. The enzyme has been well characterized with respect to its specificity, kinetics of action, amino acid sequence, and three-dimensional structure.

Chymotrypsin and a number of other enzymes are inactivated when they react with the reagent diisopropylphosphofluoridate, an extremely toxic nerve poison, capable of inhibiting acetylcholinesterase, an enzyme of the central nervous system. The reaction of reagent with enzyme is stoichiometric, with the formation of a stable diisopropylphosphoryl derivative [Eq. (11-1)].

$$\mathrm{ChT} + \mathrm{F{-}\overset{\overset{\displaystyle O}{\|}}{P}{-}}\left(\mathrm{O{-}CH}(\mathrm{CH_3})_2\right)_2 \xrightarrow{\text{pH 8}} \mathrm{ChT{-}\overset{\overset{\displaystyle O}{\|}}{P}{-}}\left(\mathrm{O{-}CH}(\mathrm{CH_3})_2\right)_2 + \mathrm{HF} \qquad (11\text{-}1)$$

Chymotrypsin + diisopropylphosphofluoridate

Diisopropylphosphoryl chymotrypsin (DIP–ChT)

The site of modification is one of the 27 serine residues of chymotrypsin, the one located at position number 195.[1] Since inactivation occurs only with the active form of the enzyme and neither with denatured enzyme molecules nor with inactive zymogen, the reactive serine 195 may be considered a component of the active site. Further support for this assignment is found in the decrease in the rate of inactivation when a chymotrypsin substrate that competes with diisopropylphosphofluoridate for the active site is present.

Additional evidence implicating serine 195 directly as an active-site component is seen in the elegant studies of Koshland and his co-workers with their chemical modification of the serine of the active site. Chymotrypsin is inactivated when treated with *p*-toluenesulfonyl fluoride, in a reaction similar to that of diisopropylphosphofluoridate [Eq. (11-2)].

$$\underset{\text{of chymotrypsin}}{\underset{\text{Serine-195}}{\overset{\displaystyle CH_2OH}{\overset{|}{-NHCHCO-}}}} + \underset{\text{fluoride}}{\underset{p\text{-toluenesulfonyl}}{F-SO_2-C_6H_4-CH_3}} \rightarrow \underset{\text{Tosyl chymotrypsin}}{\overset{\displaystyle CH_2-O-SO_2-C_6H_4-CH_3}{\overset{|}{-NHCHCO-}}} \qquad (11\text{-}2)$$

The reaction proceeds with a stoichiometry of 1 tosyl group incorporated per molecule of chymotrypsin, and chemical analysis suggests that a single serine residue is tosylated. Tosyl chymotrypsin, when treated with 0.1 *M* sodium hydroxide at 0°, undergoes a base-elimination reaction, with the formation of anhydrochymotrypsin [Eq. (11-3)].

$$\underset{\text{Tosyl chymotrypsin}}{\overset{\displaystyle CH_2-O-SO_2-C_6H_4-CH_3}{\overset{|}{-NHCHCO-}}} \rightarrow \underset{\text{Anhydrochymotrypsin}}{\overset{\displaystyle CH_2}{\overset{\|}{-NHCHCO-}}} + {}^{(-)}O_3S-C_6H_4-CH_3 \qquad (11\text{-}3)$$

The conditions of the reaction are sufficiently mild to maintain the protein in its native conformation; the only chemical change noted is the conversion of serine 195 to a residue of dehydroalanine. Anhydrochymotrypsin is inert, inactive as an enzyme catalyst, but it still binds substrates (ES-complex formation) and inhibitors. The loss of catalytic activity clearly shows that serine 195 plays an essential role in the catalytic function of the enzyme, as a component of the active site rather than of the specificity site. The small change of structure in anhydrochymotrypsin eliminates the possibility of steric hindrance of the active-site region.

[1] The numbering of amino acid residues refers to the sequence of bovine chymotrypsinogen A.

Active-site-directed reagents

Further studies on the chemical modification of the active site of chymotrypsin take advantage of an "active-site-directed reagent," one designed to form a covalent bond with the enzyme using a reaction directed by the specificity characteristics of the enzyme. Instead of using the typical substrate, *N*-tosylphenylalanine ethyl ester, which would be hydrolyzed by the enzyme, Schoellmann and Shaw prepared a substrate analog, *N*-tosylphenylalanine chloromethyl ketone, TPCK, which could serve as an inhibitor of the active site (Fig. 11-2). The sub-

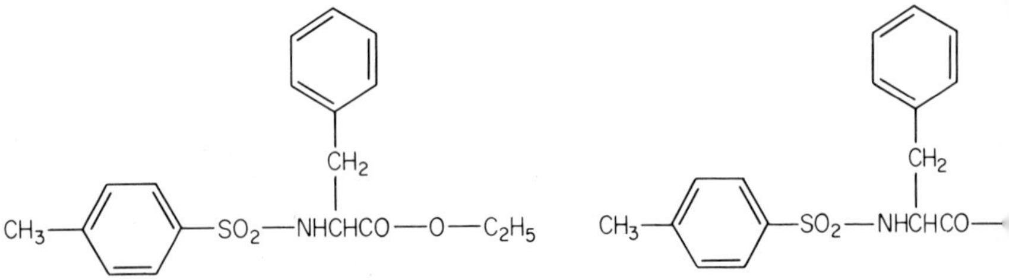

N*-*p-toluenesulfonyl-L-phenylalanyl ethyl ester

N*-*p-toluenesulfonyl-L-phenylalanine chloromethyl ketone (TPCK)

Fig. 11-2. The structure of a typical substrate of chymotrypsin (on the left) and of an active-site-directed reagent (on the right). The *p*-toluenesulfonyl moiety is referred to as tosyl (Ts) in the text.

strate analog meets the specificity requirements of the enzyme for the phenylalanyl group, but it lacks a peptide bond that could be hydrolyzed. In addition, the analog contains a chlorine atom as a chemically reactive group. The substrate analog was designed to react with the active site, by displacement of the halide and with the concommittant formation of a stable covalent bond between the *N*-tosylphenylalanine methyl ketone and the enzyme.

TPCK reacts with chymotrypsin with a 1-to-1 stoichiometry under conditions normally associated with the catalytic activity of the enzyme. Only the native enzyme participates in the reaction; chymotrypsinogen, denatured chymotrypsin, the diisopropylphosphoryl derivative, and anhydrochymotrypsin are inert. The reactive group was identified as the imidazole of one of the two histidine residues of chymotrypsin, residue number 57 of the amino acid sequence [Eq. (11-4)].

From these chemical modification studies of chymotrypsin, it is clear that both the serine and histidine participate in catalysis and, therefore, both are

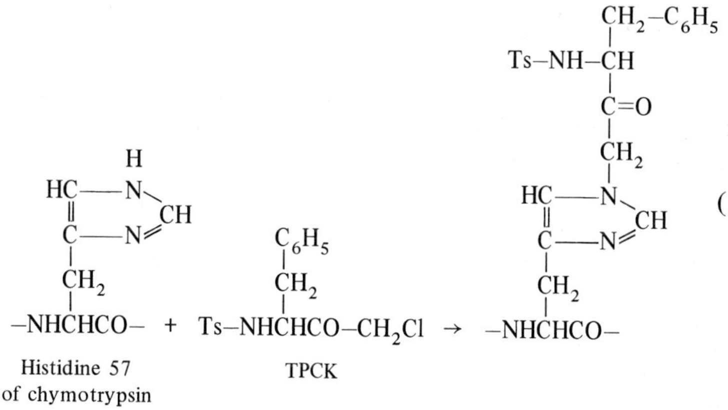

(11-4)

constituents of the active site. Furthermore, derivatives of serine 195 prevent chemical modifications of histidine 57, and derivatives of the histidine protect the serine residue. It has been postulated that the reactivity of serine 195 and histidine 57 and their role in catalysts are derived from some type of interaction of the alcohol group of serine with the imidazole of histidine. As we shall see shortly, the two residues are thought to be part of a hydrogen-bonded system.

Serine proteinases

Chymotrypsin A is only one of several proteolytic enzymes subject to inactivation by diisopropylphosphofluoridate. The closely related enzymes chymotrypsin B, trypsin, and elastase—enzymes that are elaborated by the pancreas in the form of inactive zymogens, or precursors—also react with the reagent, forming inactive diisopropylphosphoryl derivatives. A specific serine residue is the site of reaction in each case. Active-site-directed reagents, such as *N*-tosyl lysine chloromethyl ketone (Fig. 11-3), which inhibit trypsin, also implicate histidine

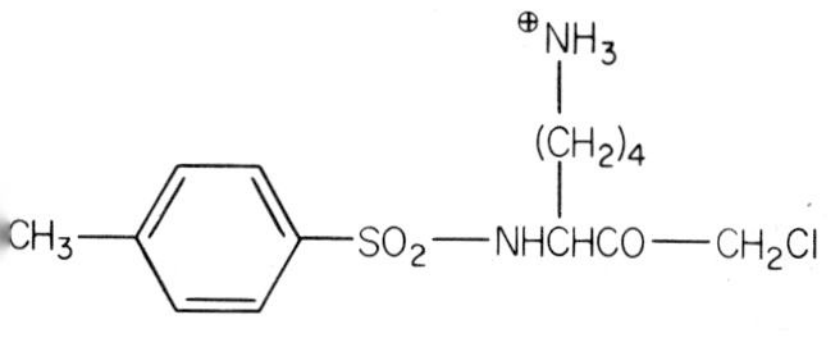

Fig. 11-3. Structure of *N*-tosyl-*L*-lysine chloromethyl ketone, an active-site-directed reagent for the enzyme trypsin. The specificity requirement of trypsin is met by the lysyl side chain of the substrate analog.

as a component of the active site. The neighboring residues in the sequence around the reactive serine and histidine are identical (Fig. 11-4). The presence of the same active-site residues in a number of different enzymes is sufficient to describe a "family" of similar proteins that must have similarities in their mechanism of enzyme action. The name "serine proteinases" refers to proteolytic enzymes with a reactive serine residue at the active site.

Active site Serine

Trypsin (bovine)	Asp · Ser · CyS · Gln · Gly · Asp · Ser · Gly · Gly · Pro · Val · Val · Cys · Ser · Gly · Lys
Chymotrypsin *A* (bovine)	Ser · Ser · CyS · Met · Gly · Asp · Ser · Gly · Gly · Pro · Leu · Val · Cys · Lys
Chymotrypsin *B* (bovine)	Ser · Ser · Cys · Met · Gly · Asp · Ser · Gly · Gly · Pro · Leu · Val · Cys · Gln
Elastase (porcine)	Ser · Gly · Cys · Gln · Gly · Asp · Ser · Gly · Gly · Pro · Leu · His · CyS · Leu
Thrombin (bovine)	Asp · Ala · CyS · Glu · Gly · Asp · Ser · Gly · Gly · Pro · Phe · Val · Met · Lys

Active site Histidine

Trypsin (bovine)	Val · Val · Ser · Ala · Ala · His · CyS · Tyr · Lys
Chymotrypsin *A* (bovine)	Val · Val · Thr · Ala · Ala · His · Cys · Gly · Val
Chymotrypsin *B* (bovine)	Val · Val · Thr · Ala · Ala · His · Cys · Gly · Val
Elastase (porcine)	Val · Met · Thr · Ala · Ala · His · CyS · Val · Asp
Thrombin (bovine)	Thr · Ala · Ala · His · CyS

Fig. 11-4. The amino acid sequences of a family of proteolytic enzymes in the immediate region of the active-site serine and histidine.

Active-site residues of hydrolytic enzymes

Chemical-modification techniques for the identification of active-site residues of hydrolytic enzymes showed that only a limited number of different types of amino acid residues are present as active-site components (Table 11-1). In addition to the serine-histidine of serine proteinases, other "families" of enzymes have cysteine-histidine, histidine-histidine, and glutamic acid-aspartic acid as active-site constituents. It is obvious that the list may be extended as the active sites of other enzymes are identified.

TABLE 11-1

Active-Site Residues

Enzyme	*Amino Acid Residues*[a]
Chymotrypsin[b]	Ser 195-His 57-Asp 102
Papain	CySH 25-His 159-Asn 175
Ribonuclease	His 12-His 119
Lysozyme	Glu 35-Asp 52
Carboxypeptidase A	Glu 270-Tyr 248

[a]The amino acids believed to be components of the active site of several enzymes of known sequence and of known x-ray structure.

[b]The residue numbers refer to the sequence of bovine chymotrypsinogen A.

Serine 195 and histidine 57, which are widely separated residues in the amino acid sequence of chymotrypsin, nevertheless must be close to one another in the three-dimensional structure. As components of the active site, the side chains of serine 195 and histidine 57 lie close to the susceptible peptide bond of substrate, and they must participate in the bond-cleavage events that are a part of the catalytic action of the enzyme.

The complete three-dimensional structure of chymotrypsin is now known, thanks to the efforts of Blow on the x-ray crystallography and to Hartley, who established the amino acid sequence. Consideration of the three-dimensional structure is necessary for a better appreciation of the arrangement of the groups at the active center and an understanding of the mechanism of action.

The three-dimensional structure

Before proceeding directly to the x-ray crystallographic studies of chymotrypsin, it should be recalled that the analysis of protein structure at high resolution was first applied to the heme proteins, myoglobin and hemoglobin (Chap. 1). The detailed information on the secondary and tertiary structure of myoglobin provided the necessary incentive for studies of other types of proteins.

Soon after the structural studies of the heme proteins were completed, Phillips and his co-workers reported on the x-ray analysis of lysozyme, an enzyme isolated from hen egg-white. These investigators obtained the structure of the enzyme and also of a crystalline enzyme-inhibitor complex. The lysozyme-inhibitor complex is similar to an enzyme-substrate complex in that interactions existing between large and small molecules resemble the interactions between the enzyme and substrate. Thus the structural features of the enzyme concerned with specificity characteristics (the specificity site) may be deduced from the study of the complex. Then the active site (catalytic groups) may be tentatively identified from a knowledge of the bond susceptible to hydrolysis in the substrate and the presence of functional groups of the enzyme close to the susceptible bond.

X-ray studies of chymotrypsin

Following closely behind these studies on lysozyme, Blow and his group tried a similar approach with the proteolytic enzyme chymotrypsin. A tosyl derivative of bovine chymotrypsin A was prepared from the reaction of the active form of the enzyme with *p*-toluenesulfonyl fluoride, which produces an inactive derivative of the active-site serine 195 [Eq. (11-2)]. The x-ray analysis of the inactive enzyme gave diffraction patterns suitable for the solution of the complete structure of the protein.

Serine 195 was easily identified in the electron-density maps as its tosyl derivative. Serine 195 is on the surface of the molecule, at the top of a small crevice, and located close to the middle of the spheroidal structure (Fig. 11-5). Histidine 57 is close to the serine, so that the oxygen of the serine alcohol and the

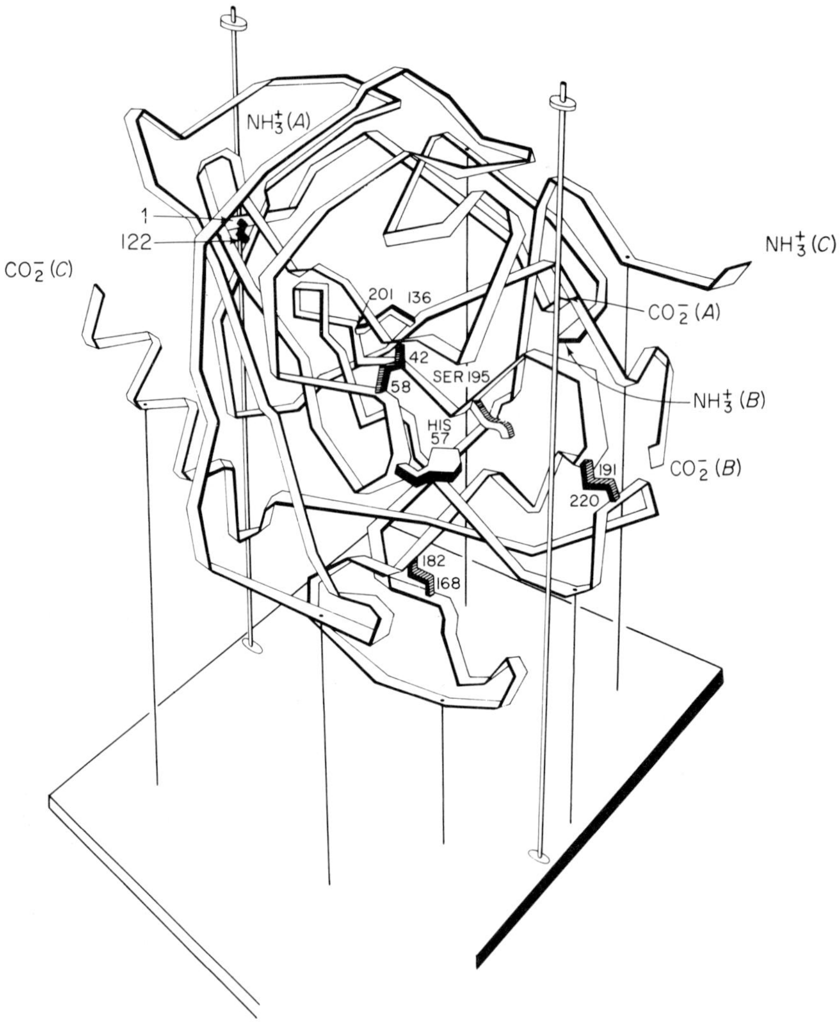

Fig. 11-5. A schematic drawing of the backbone structure of α-chymotrypsin with the active site residues indicated. [Adapted from Sigler, et al., *J. Mol. Biol.*, **35**, 143 (1968).]

nitrogen (Nε2) of the imidazole are within hydrogen-bonding distance of one another (Fig. 11-6).[2] The remainder of the imidazole ring is partly buried with

[2]The designation Nδ1 refers to the nitrogen atom of the imidazole ring closest to the alpha carbon while the Nε2 nitrogen atom is farthest away. The abbreviations and symbols used for the description of the conformation of polypeptide chains are based on an IUPAC-IUB Commission on Biochemical Nomenclature [see *J. Biol. Chem.*, **245**, 6489 (1970)].

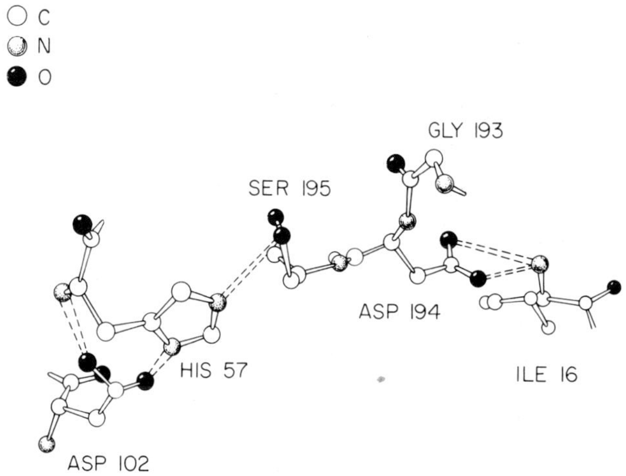

Fig. 11-6. The conformation of a few amino acids in the active center of α-chymotrypsin. Hydrogen bonds are dashed lines. [From D. M. Blow and T. A. Steitz, *Ann. Rev. Biochem.*, 39, 63 (1970).]

the second nitrogen atom (Nδ1) close to a buried carboxylate group of aspartic acid 102.

The charge-relay system

It is rather surprising that the carboxyl group of aspartyl 102 is ionized because the functional group is buried within the molecule and located in a hydrophobic environment. Blow recognized the potential for hydrogen bonding between the buried carboxylate and the neighboring NH of the imidazole (nitrogen atom Nδ1). These observations suggested to him that the active site may consist of a hydrogen-bonded network, or charge-relay system, of the three residues Ser 195-His 57-Asp 102. The charge-relay system would promote a proton transfer from serine 195 to histidine 57, thereby increasing the nucleophilicity of the serine oxygen atom (Fig. 11-7).

The serine oxygen atom as a nucleophile attacks the carbonyl of the peptide bond of appropriate substrates leading to the formation of an acyl-enzyme intermediate (Fig. 11-8). The acyl-enzyme can be hydrolyzed by a water molecule with assistance from histidine 57, which participates in these reactions as a general acid-general base. The mechanism of action is consistent with the kinetic behavior of the enzyme, and the charge-relay system provides a reasonable explanation for the unique reactivity of serine 195 as an active site component and in the chemical-modification studies discussed earlier.

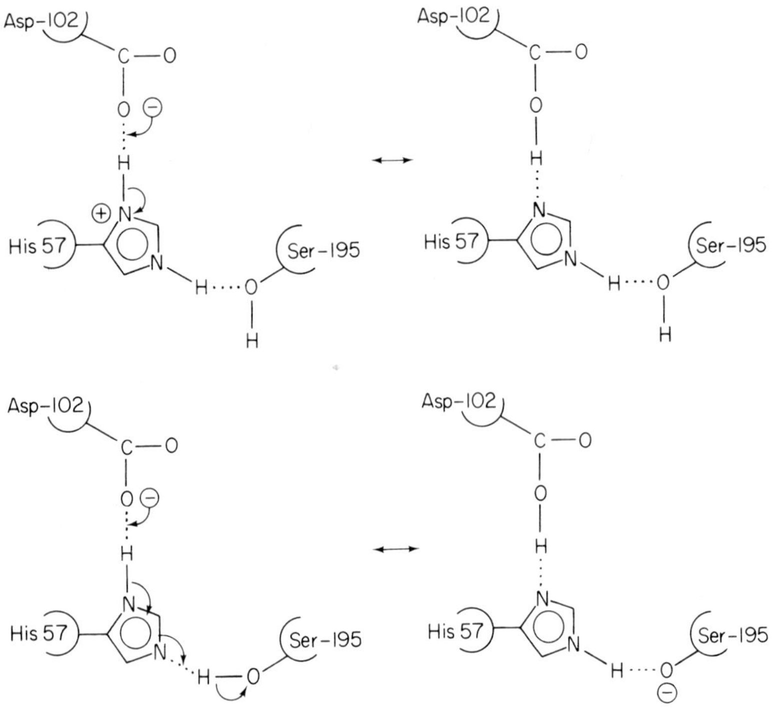

Fig. 11-7. Canonical forms of the active center of α-chymotrypsin; top line, pH 4; bottom line, pH 8. The forms represented indicate purely electronic rearrangements. The protons in hydrogen bonds are represented schematically as being near the atom to which they are covalently bonded in the more "conventional" of the canonical forms, namely, the forms on the left. The forms on the right are drawn with "stretched" electronic bonds to these protons. [From D. M. Blow, J. J. Birktoft, and B. S. Hartley, *Nature,* **221**, 337 (1969).]

The specificity site

X-ray studies also provided definitive information on the specificity site of chymotrypsin. This region was recognized in the electron-density maps as the structure binding the tosyl group, a reasonable assumption based on the known specificity of the enzyme for aromatic structures. The tosyl group is located in a pocket of the enzyme surface, referred to as the "tosyl hole," in which the aromatic group interacts with hydrophobic groups lining the pocket. The size of the pocket can readily accommodate side chain substituents of normal substrates. A bound substrate is then correctly oriented on the enzyme so that the susceptible peptide bond is within bonding distance of the active-site serine.

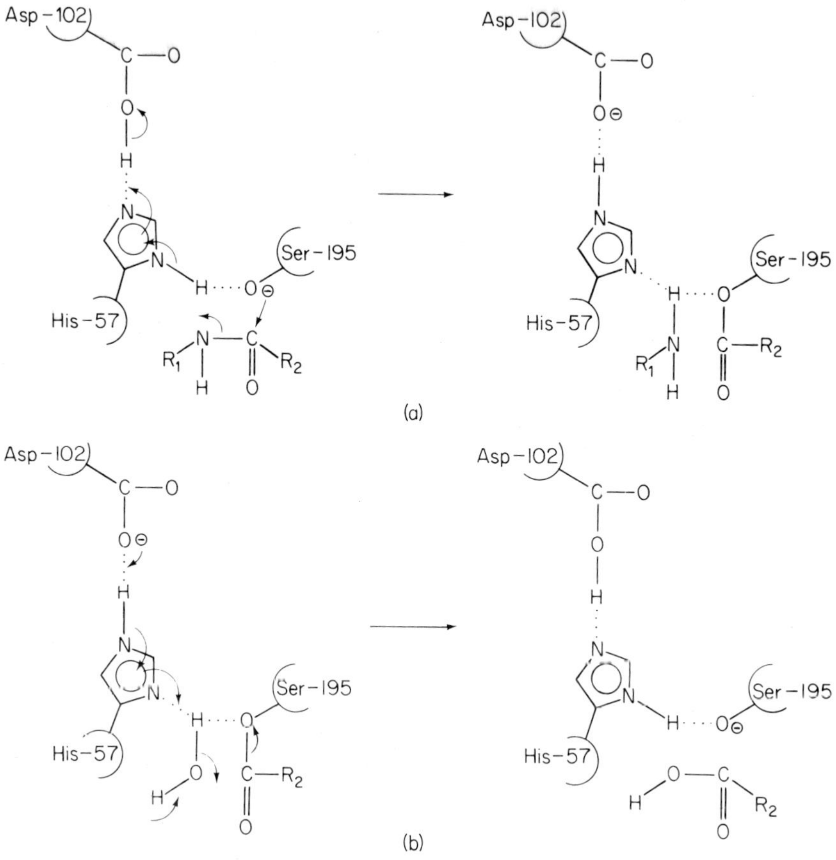

Fig. 11-8. A proposed mechanism of action of α-chymotrypsin for the hydrolysis of an ester substrate. [From D. M. Blow and T. A. Steitz, *Ann. Rev. Biochem.*, **39**, 63 (1970).]

Further studies on the x-ray diffraction patterns of *N*-formyl tryptophan bound to chymotrypsin confirmed this idea. *N*-formyl tryptophan is a virtual substrate, a product of the hydrolysis of a corresponding ester or amide of *N*-formyl tryptophan, and it is presumed to bind to the enzyme in the same mode as a normal substrate undergoing hydrolysis. The "tosyl hole" is indeed the substrate binding site, accommodating the indole group quite satisfactorily, while the tryptophanyl carboxyl group is probably making contact with both serine 195 and histidine 57. In summary, these elegant studies of enzyme-inhibitor complexes increased our knowledge of the active center of the enzyme and led to a reasonable proposal of the mechanism of action in the hydrolysis of both small and large substrates.

AMINO ACID SEQUENCE COMPARISONS

Earlier in this chapter it was noted that the serine-histidine interaction at the active site of chymotrypsin is also a feature found in other enzymes. With the elucidation of the amino acid sequence of several of these enzymes, it became possible to compare one with another and to look for additional similarities in their structures (Fig. 11-9). The sequences of proteins, which may differ in the total number of residues, are compared by aligning each sequence in a series of parallel lines. The residues are aligned throughout the sequence to maximize

	1	2	3	4	5	6	7	8	9	10	11	12	13	14	15	16
C	Cys	Gly	Val	Pro	Ala	Ile	Gln	Pro	Val	Leu	Ser	Gly	Leu	Ser	ARG	Ile
T										Val	Asp	Asp	Asp	Asp	LYS	Ile
E																VAL

	17	18	19	20	21	22	23	24	25	26	27	28	29	30	31	32
C	Val	Asn	Gly	Glu	Glu	Ala	Val	Pro	Gly	Ser	Trp	Pro	Trp	Gln	Val	Ser
T	Val	Gly	Gly	Tyr	Thr	Cys	Gly	Ala	Asn	THR	Val	Pro	TYR	Gln	Val	Ser
E	Val	Gly	Gly	Thr	Glu	Ala	Gln	Arg	Asn	Ser	Trp	Pro	Ser	Gln	ILE	Ser

	33	34	35	36	36A	36B	36C	37	38	39	40	41	42	43	44	45
C	Leu	Gln	Asp	LYS	—	—	—	THR	Gly	PHE	His	Phe	Cys	Gly	Gly	Ser
T	Leu	ASN	—	—	—	—	—	Ser	Gly	TYR	His	Phe	Cys	Gly	Gly	Ser
E	Leu	Gln	Tyr	Arg	Ser	Gly	Ser	Ser	Trp	Ala	His	Thr	Cys	Gly	Gly	THR

	46	47	48	49	50	51	52	53	54	55	56	57	58	59	60	61
C	Leu	Ile	Asn	GLU	Asn	Trp	Val	Val	Thr	Ala	Ala	His	Cys	Gly	Val	THR
T	Leu	Ile	Asn	Ser	GLN	Trp	Val	Val	SER	Ala	Ala	His	Cys	Tyr	Lys	SER
E	Leu	Ile	Arg	GLN	Asn	Trp	Val	Met	Thr	Ala	Ala	His	Cys	Val	Asp	Arg

	62	63	64	65	65A	66	67	68	69	70	71	72	73	74	75	76
C	Thr	Ser	Asp	Val	—	Val	Val	Ala	Gly	Glu	Phe	ASP	Gln	Gly	Ser	Ser
T	Gly	ILE	Gln	Val	Arg	Leu	—	—	Gly	GLN	Asp	Asn	ILE	Asn	Val	Val
E	Glu	LEU	Thr	Phe	Arg	Val	Val	VAL	Gly	Glu	His	Asn	LEU	Asn	Gln	Asn

	77	78	79	80	81	82	83	84	85	86	87	88	89	90	91	92
C	Ser	Glu	Lys	Ile	Gln	Lys	LEU	Lys	ILE	Ala	Lys	VAL	Phe	Lys	Asn	Ser
T	Glu	Gly	Asn	GLN	Gln	PHE	Ile	Ser	Ala	Ser	Lys	Ser	Ile	Val	His	Pro
E	Asn	Gly	Thr	Glu	Gln	TYR	VAL	Gly	VAL	Gln	Lys	ILE	VAL	Val	His	Pro

	93	94	95	96	97	98	99	99A	99B	100	101	102	103	104	105	106
C	Lys	Tyr	Asn	Ser	Leu	Thr	ILE	—	—	Asn	Asn	Asp	Ile	Thr	Leu	Leu
T	Ser	Tyr	Asn	Ser	ASN	Thr	LEU	—	—	Asn	Asn	Asp	Ile	Met	Leu	ILE
E	Tyr	TRP	Asn	THR	ASP	Asp	Val	Ala	Ala	Gly	Tyr	Asp	Ile	Ala	Leu	Leu

	107	108	109	110	111	112	113	114	115	116	117	118	119	120	121	122
C	Lys	Leu	Ser	Thr	Ala	Ala	Ser	Phe	Ser	Gln	Thr	Val	Ser	Ala	VAL	Cys
T	Lys	Leu	Lys	Ser	Ala	Ala	Ser	Leu	Asn	Ser	Arg	Val	Ala	Ser	ILE	Ser
E	ARG	Leu	Ala	Gln	Ser	Val	THR	Leu	Asn	Ser	Tyr	Val	Gln	Leu	Gly	Val

Fig. 11-9. The complete amino acid sequences of bovine chymotrypsinogen A (top line, letter C), bovine trypsinogen (middle line, letter T), and porcine elastase (bottom line, letter E). The numbering of

similarities. The introduction of a dash is used when gaps are found in the sequence because the proteins differ in the total number of amino acid residues.

Sequences of trypsin and chymotrypsin

If the sequences of bovine trypsin and bovine chymotrypsin A are compared, almost 40 percent of the residues are identical (Fig. 11-9). Particularly noteworthy is the identity of the sequences in the immediate region surrounding

	123 124 125 126 127 128 129 130 131 132 133 134 135 136 137 138
C	Leu–Pro–SER–Ala–Ser–Asp–Asp–Phe–Ala–Ala–Gly–Thr–Thr–Cys–Val–Thr
T	Leu–Pro–THR — Ser–Cys–Ala–Ser — Ala–Gly–Thr–Gln–Cys–Leu–Ile
E	Leu–Pro–Arg–Ala–Gly–Thr–Ile–Leu–Ala–Asn–Asn–SER–Pro–Cys–Tyr–Ile

	139 140 141 142 143 144 145 146 147 148 149 150 151 152 153 154
C	Thr–Gly–Trp–Gly–Leu–Thr–Arg–Tyr–Thr–Asn–Ala–ASN–Thr–Pro–Asp–Arg
T	SER–Gly–Trp–Gly–Asn–Thr–LYS–Ser–SER–Gly–Thr–Ser–Tyr–Pro–Asp–Val
E	Thr–Gly–Trp–Gly–Leu–Thr–Arg — Thr–Asn–Gly–GLN–Leu–Ala–GLN–Thr

	155 156 157 158 159 160 161 162 163 164 165 166 167 168 169 170
C	Leu–Gln–Gln–Ala–Ser–Leu–Pro–LEU–Leu–Ser–Asn–THR–Asn–Cys–Lys–Lys
T	Leu–Lys–Cys–Leu–Lys–Ala–Pro–ILE–Leu–Ser–Asn–SER–Ser–Cys–Lys–Ser
E	Leu–Gln–Gln–Ala–Tyr–Leu–Pro–Thr–Val–Asp–Tyr–Ala–Ile–Cys–Ser–Ser

	170A 170B 171 172 173 174 175 176 177 178 179 180 181 182 183 184
C	— — Tyr–Trp–Gly–THR–LYS–Ile–Lys–ASP–Ala–Met–ILE–Cys–Ala–Gly
T	— — Ala–TYR–Pro–Gly–Gln–Ile–Thr–Ser–Asn–Met–Phe–Cys–Ala–Gly
E	Ser–Ser–Tyr–Trp–Gly–SER–Thr–Val–Lys–ASN–Ser–Met–VAL–Cys–Ala–Gly

	184A 185 186 187 187B 188 188A 189 190 191 192 193 194 195 196 197
C	— Ala–Ser–Gly — Val — Ser–Ser–Cys–Met–Gly–Asp–Ser–Gly–Gly
T	Tyr–Leu–Glu–Gly–Gly–Lys — Asp–Ser–Cys–Gln–Gly–Asp–Ser–Gly–Gly
E	— Gly–Asn–Gly — Val–Arg–Ser–Gly–Cys–Gln–Gly–Asp–Ser–Gly–Gly

	198 199 200 201 202 203 204 205 206 207 208 209 210 211 212 213
C	Pro–Leu–Val–Cys–Lys–Lys–Asn–Gly–Ala–Trp–Thr–Leu–Val–Gly–Ile–Val
T	Pro–Val–Val–Cys–Ser–Gly–Lys — — — — Leu–Gln–Gly–Ile–Val
E	Pro–Leu–His–Cys–Leu–Val–Asn–Gly–Gln–TYR–Ala–VAL–His–Gly–VAL–Thr

	214 215 216 217 217A 218 219 220 221 221A 222 223 224 225 226 227 228 229
C	Ser–Trp–Gly–Ser — Ser–Thr–Cys–Ser — Thr–Ser–Thr–Pro–Gly–Val–Tyr–Ala
T	Ser–Trp Gly Ser — — Gly–Cys–Ala–Gln–Lys–Asn–Lys–Pro–Gly–Val–Tyr–Thr
E	Ser–PHE–Val–Ser–Arg–Leu–Gly–Cys–Asn–Val–Thr–Arg–Lys–Pro–Thr–Val–PHE–Thr

	230 231 232 233 234 235 236 237 238 239 240 241 242 243 244 245
C	Arg–Val–THR–Ala–Leu–Val–Asn–Trp–VAL–Gln–GLN–Thr–LEU–Ala–Ala–Asn
T	LYS–Val–Cys–Asn–Tyr–Val–Ser–Trp–Ile–Lys–Gln–Thr–Ile–Ala–Ser–Asn
E	Arg–Val–SER–Ala–Tyr–ILE–Ser–Trp–Ile–ASN–ASN–Val–Ile–Ala–Ser–Asn

residues is based on the sequence of chymotrypsinogen. Residues which are chemically similar (Lys-Arg, etc.) are shown in *capitals.* Identical residues are in *bold-face.*

the active-site residues—the serine 195, histidine 57, and aspartic acid 102.

Throughout the sequence of trypsin and chymotrypsin, there are many regions of identical residues. In addition, there are many conservative substitutions where one type of chemical grouping is substituted for another of a similar type. There are frequent interchanges of lysine and arginine, glutamine and asparagine, serine and threonine, isoleucine and leucine and valine. Conservative substitutions generally introduce only minor alterations in the conformation of the molecule.

Sequence homologies

The overall similarity of the sequence of trypsin and chymotrypsin, when identical and conservative substitutions are grouped together, is close to 57 percent of the structure. If a second comparison is made, that between bovine chymotrypsins A and B, two enzymes with identical mechanisms of action and almost identical specificities, the overall chemical similarity is as high as 78 percent. Thus two or more proteins with a similar biological function, with identical residues at the active site, may be closely related to one another in their total sequence (exceptions also arise and a specific case will be discussed later).

The three-dimensional structures of closely related sequences must be similar to one another. The similarities are a consequence of the fact that folding of the polypeptide chain is dictated by the amino acid sequence. Recent studies of electron-density maps of crystals of trypsin and elastase indicate a closeness of their structures to that of chymotrypsin. Where differences exist in the amino acid sequence, these residues are usually found on the surface of the molecule or at bends of the polypeptide chain. At both regions, a change from one residue to another or the addition of extra residues can be readily accommodated with a minimal change in the total effect on conformation.

COMMENTS ON ENZYME STRUCTURE AND FUNCTION

At this point it may be helpful if we summarize several characteristic properties of enzymes that stress the relationship between structure and function. First, biological activity is dependent on an intact structure, that of the native enzyme, in which the molecule maintains a specific three-dimensional arrangement. The macromolecular structure is so designed that it binds substrate molecules in a specific mode—a feature responsible for the selection of an appropriate molecule as substrate, and for the proper orientation of the substrate so that the enzyme participates in productive catalytic changes (Table 11-2). Catalytic activity is derived from a few residues comprising the active site, a region of the enzyme located on the surface in close proximity to the bound substrate molecule. The active-site functional groups participate in the chemical events

TABLE 11-2

Catalytic Effects Revealed by Diffraction Studies of α-Chymotrypsin[a]

	Catalytic Effect	*Relevant Data from X-Ray Diffraction of Chymotrypsin*
Proximity	Increases concentration of substrate at active site	Substrate binding pocket CO(Ser 214)···HN(Substrate) H bond
Orientation	Restricts possible conformations of ES complex to productive mode	Same
Specificity	Restricts types of substrate able to bind in productive mode	Same
Induced fit	Substrate binding changes enzyme conformation to a productive one	Not believed to occur in chymotrypsin
Substrate strain	Substrate distorted energetically toward its transition state	Not detectable from present results, but H bonds to substrate CO from NH(193) and NH(195) could contribute
Enzyme strain	Energy stored in enzyme reduces activation energy of transition	Asp 102···His 57···Ser 195 charge-relay system
Chemical catalysis	General acid, general base, nucleophilic, etc.	Polarized H bond between His 57 and Ser 195 presumably relevant
Chemical intermediates	Separate reaction steps of lower activation energy than overall process	Structure of a stable acyl-enzyme (indoleacryolyl-)

[a]Adapted from R. Henderson, C. S. Wright, G. P. Hess, and D. M. Blow, *Cold Spring Harbor Symposia on Quantitative Biology,* **36**, 63 (1972).

that occur when the substrate is converted to products. Thus, the arrangement of the active site is designed in such a way as to enable amino acid residues to participate catalytically as nucleophiles, electrophiles, and as general acids and general bases.

The active center is absolutely essential as a structural region of an enzyme, and its structural features are responsible for the remarkable specificity characteristics and high-catalytic efficiency that make enzymes behave so uniquely as catalysts. But it is clear that only a small part of an enzyme's structure contributes to the formation of the active center. The remainder of the molecule is apparently necessary as a frame, or structural support, so that residues of the active center are maintained in an appropriate three-dimensional arrangement.

PROTEIN STRUCTURE AND EVOLUTION

We now turn our attention to a new subject, namely an examination of the amino acid sequence of a protein as it varies with the biological source. The protein we will examine as a typical case is cytochrome c, an essential protein component of the electron-transport system of aerobic organisms.

Structure of Cytochrome c

Cytochrome c is a hemeprotein consisting of a single polypeptide chain of 104 amino acid residues and a heme group covalently linked to the side chains of two amino acid residues (Fig. 11-10). In a series of elegant studies performed

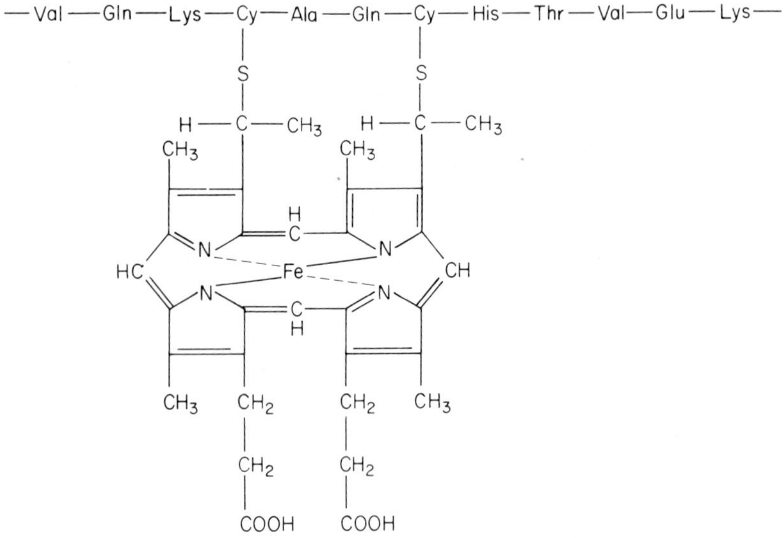

Fig. 11-10. The structure of the heme group of beef heart cytochrome c and the covalent linkage to the polypeptide chain. The partial sequence, as shown is called the heme peptide. The structure of the heme group is identical in all cytochromes c but the sequence varies.

by Margoliash and by Smith, the structure of cytochrome c was established from each of 30 different species, varying from the cytochrome c of man to that of yeast (Fig. 11-11). Using in vitro methods to examine biological activity, the cytochromes of the 30 species behaved in the same way and were equally effective as electron carriers in a mitochondrial system. Thus cytochrome c structures offer an excellent opportunity to examine the effect of variable sequences

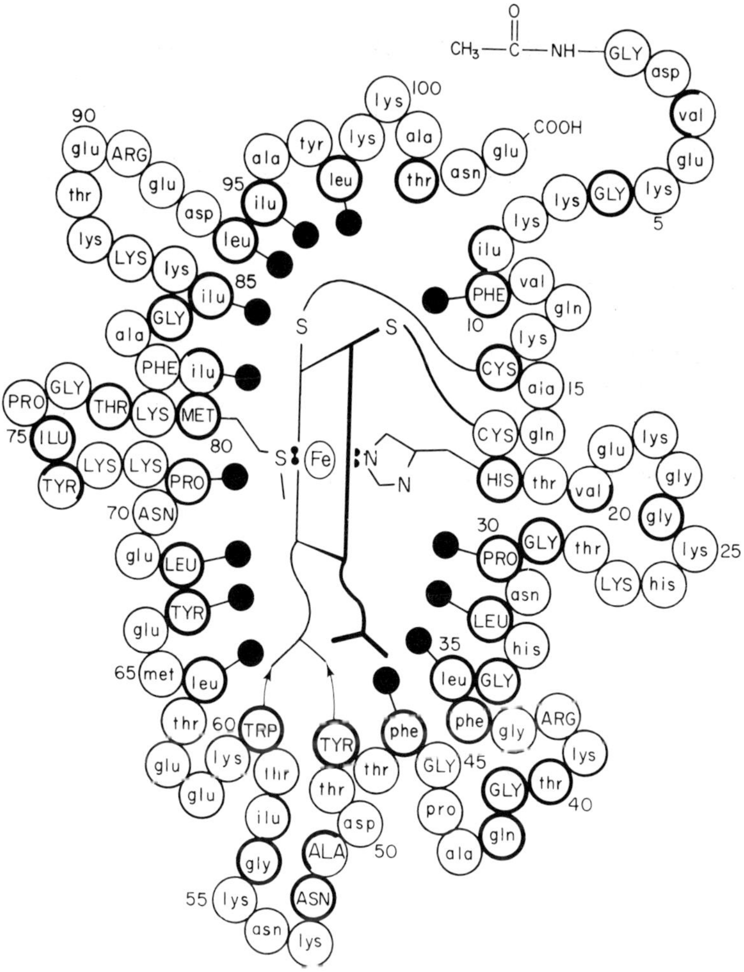

Fig. 11-11. The amino acid sequence and the heme environment of horse heart cytochrome c. *Heavy circles* indicate side chains that are buried within the interior of the molecule. *Light circles* indicate side chains that are on the outside of the molecule. *Dark half-circles* are used for those side chains that are half-buried at the surface. Those residues with side chains packing against the heme group are indicated by the addition of a large *black dot*. *Arrows* from tryptophan 59 and tyrosine 48 to the buried propionic acid group represent hydrogen bonds. The assignments of the residues in space are based on x-ray studies at 2.8 Å resolution. Amino acid residues designated by *capital letters* are totally invariant in the sequences of cytochromes c in 30 species. [From R. E. Dickerson, T. Takano, D. Eisenberg, O. B. Kallai, L. Samson, A. Cooper, and E. Margoliash, *J. Biol. Chem.*, **246**, 1511 (1971).]

(species variation) on the three-dimensional structure while the biological activity remains essentially constant.

The amino acid sequences of cytochrome c are remarkably alike even when comparisons include distantly related species. All vertebrate cytochromes c have 104 amino acids with an *N*-acetylated amino-terminal glycyl residue and extensive sequence homologies throughout. However, invertebrate, plant, and fungal cytochromes c include eight or more additional residues, with these extra residues occurring as extensions of the sequence at the amino-terminal end. When the entire sequences are aligned, the extra residues do not influence the sequence homology existing in the remainder of the molecule. Based on extensive homology of the entire sequence, it is clear that the extra amino-terminal residues must be nonessential for both stability and function.

In all structures of cytochrome c, the heme group is covalently linked to the polypeptide chain, while in the case of hemoglobin and myoglobin the heme group is bound noncovalently. The covalent attachment occurs at cysteine residues 14 and 17 by a thioether link to the vinyl groups of the heme. The site of attachment is close to the amino-terminal end of the polypeptide chain, and this feature is found in all cytochromes c[3] (Fig. 11-11).

The iron atom is found in cytochrome c in an octahedral configuration. Four ligands of the iron are the nitrogens of the cyclic tetrapyrrole, and two are contributed by amino acid residues of the polypeptide. The amino acid ligands are the imidazole nitrogen of histidine 18 and the sulfur atom of methionine 80 (Fig. 11-11).

Histidine 18 and methionine 80 are invariant residues in the sequences of all cytochromes c studied so far, suggesting that both ligands may be essential for a cytochrome c to be biologically functional. In its biological role, cytochrome c participates in the electron-transport system with the valence of the iron atom changing reversibly between the ferric form (oxidized state) and the ferrous form (reduced state).

The x-ray diffraction studies of cytochrome c as reported by Dickerson have increased our appreciation of the general shape of the molecule and clarified the role of individual residues as structural elements. The conformation of the polypeptide chain forms a crevice for the heme group; residues 1 to 47 lie on one side of the heme and residues 48 to 91 on the other. The crevice itself is lined by 16 hydrophobic residues (Fig. 11-11) and includes the side chains of leucine (6, in number), isoleucine (3), phenylalanine (2), proline (2), tyrosine (2), and tryptophan (1). Since the structure of the crevice appears to be critical in

[3]An interesting exception was noted in the amino acid sequence of cytochrome c from *Euglena gracilis*. A single linkage of the heme to protein involves a thioether bond with half-cystine residue 17. At position 14, alanine substitutes for cysteine. [See *Nature*, **241**, 531-535 (1973).]

determining the total three-dimensional structure, the hydrophobic amino acid residues of the crevice region are mainly invariant in all cytochromes c.

Other invariant residues of cytochrome c are necessary in order to maintain a proper conformation of the molecule. Glycine or proline occurs at bends where the polypeptide chain shows a change in direction, and where these residues appear to be absolute requirements, for other amino acids could not fit at some of the turns. Since glycine lacks a side chain, it offers minimal steric hindrance in crowded regions. Proline lacks the planar peptide bond and its ring structure is easily accommodated at bends. Glycine also occurs at regions of close contact between segments of neighboring polypeptide chains.

Conservative substitutions

Many variable residues found throughout the sequence of cytochromes c are conservative substitutions. For example, there are interchanges of valine and isoleucine, leucine and isoleucine, tyrosine and phenylalanine, aspartic acid and glutamic acid, and others of a similar nature. The substitution of one residue by another with similar properties implies that the folding of the polypeptide chain and the stability of the globular structure may be satisfied by one of several amino acid residues appearing at a number of different positions in the sequence of the protein. It is also possible that these sequences of the polypeptide are not as important as structural elements as those regions containing invariant residues.

There are also many sites in the amino acid sequence that can be satisfied by one of several types of amino acids. This is clearly the case at residue 92, where nine substitutions occurred, at residue 89 with eight, and at residue 60 with seven substitutions. A number of other positions have multiple substitutions at different sites along the polypeptide chain. When substitutions occur so freely, these residues must be considered as nonessential for stability and function. Most conservative changes are located on the surface of the molecule where they may interact with solvent molecules rather than with buried residues.

As the number of known sequences of cytochrome c from different species increased, these data were used to propose a chemical basis of evolution. Before we examine individual sequences of cytochrome c, we will briefly give an elementary description of the evolutionary process from the biological point of view.

Evolutionary process

Evolution is directly related to the transmission of hereditary characteristics, typically in the way DNA molecules are transmitted from parents to progeny. Normal informational transfer exists as long as the base sequence of parental DNA molecules are accurately copied. Mutations caused by radiation and certain chemicals can alter the structure of DNA. The modified DNA, or gene segment, ultimately is responsible for a change in the chemical structure of one or more cellular components of the progeny, leading to different structural and biological properties. For example, if a mutation causes the complete or partial loss of an important biological activity, the organism often is unable to compete successfully. Natural selection in turn acts to restrict the proliferation of organisms carrying genetic defects.

Some mutations may be only somewhat deliterious in their effects while others may be responsible for improvements in a particular structure or function. Selective pressures again favor the propagation of well-adapted organisms. A direct assessment of a mutational change is seen in the altered sequence of a protein; consequently, comparative-sequence analyses provide a measure of the relatedness of surviving organisms.

All living organisms of today's world are believed to have evolved from a single ancestoral species, a primitive organism that existed at some time in the distant past. Although existing species may be highly complex compared to their ancestors, they still retain vestiges of the past. One assumes a conservative nature for the evolutionary process, and therefore, species that diverged early in the biological time scale more closely resemble their ancestors than those of more recent origin. The relatedness of amino acid sequences of a specific protein in different species thus suggests the relative order in which the proteins became distinct from one another and how the species themselves may have developed.

The sequences of cytochrome c from different species are the most extensive data presently available that can be used for an analysis of the relatedness of species. The number of amino acid differences between the cytochrome c of one species and that of another gives a rough estimate of whether the two are closely related or are far removed on the evolutionary scale (Table 11-3). For example, the cytochrome c of man differs from that of the rhesus monkey in one residue, from that of the horse in 12, the chicken in 13, the tuna in 21, and yeast in 45 residues.

Phylogenetic tree of cytochrome c

Based on a rather sophisticated genetic analysis of sequence relatedness, a phylogenetic tree may be constructed depicting the evolution of cytochrome c (Fig. 11-12). The tree shows all existing species known today at the ends of branches, and their descent from common precursors is indicated by appropriate branches from the trunk of the tree. The first organism that evolved with a gene for a functional cytochrome c at some period in the distant past is shown

TABLE 11-3

Number of Amino Acid Differences Between Sequences of Cytochrome c[a]

	Human	Rhesus monkey	Horse	Donkey	Pig, bovine, sheep	Dog	Gray whale	Rabbit	Kangaroo	Chicken and turkey	Penguin	Pekin duck	Snapping turtle	Rattlesnake	Bullfrog	Tuna fish	Screw-worm fly	Silkworm moth	Wheat	*Neurospora crassa*	Baker's yeast	*Candida krusei*
Human	0	1	12	11	10	11	10	9	10	13	13	11	15	14	18	21	27	31	43	48	45	51
Rhesus monkey	1	0	11	10	9	10	9	8	11	12	12	10	14	15	17	21	26	30	43	47	45	51
Horse	12	11	0	1	3	6	5	6	7	11	12	10	11	22	14	19	22	29	46	46	46	51
Donkey	11	10	1	0	2	5	4	5	8	10	11	9	10	21	13	18	22	28	45	46	45	50
Pig, bovine, sheep	10	9	3	2	0	3	2	4	6	9	10	8	9	20	11	17	22	27	45	46	45	50
Dog	11	10	6	5	3	0	3	5	7	10	10	8	9	21	12	18	21	25	44	46	45	49
Gray whale	10	9	5	4	2	3	0	2	6	9	9	7	8	19	11	17	22	27	44	46	45	50
Rabbit	9	8	6	5	4	5	2	0	6	8	8	6	9	18	11	17	21	26	44	46	45	50
Kangaroo	10	11	7	8	6	7	6	6	0	12	10	10	11	21	13	18	24	28	47	49	46	51
Chicken and turkey	13	12	11	10	9	10	9	8	12	0	2	3	8	19	11	17	23	28	46	47	46	51
Penguin	13	12	12	11	10	10	9	8	10	2	0	3	8	20	12	18	24	27	46	48	45	50
Pekin duck	11	10	10	9	8	8	7	6	10	3	3	0	7	17	11	17	22	27	46	46	46	51
Snapping turtle	15	14	11	10	9	9	8	9	11	8	8	7	0	22	10	18	24	28	46	49	49	53
Rattlesnake	14	15	22	21	20	21	19	18	21	19	20	17	22	0	24	26	29	31	46	47	47	51
Bullfrog	18	17	14	13	11	12	11	11	13	11	12	11	10	24	0	15	22	29	48	49	47	51
Tuna fish	21	21	19	18	17	18	17	17	18	17	18	17	18	26	15	0	24	32	49	48	47	48
Screw-worm fly	27	26	22	22	22	21	22	21	24	23	24	22	24	29	22	24	0	14	45	41	45	47
Silkworm moth	31	30	29	28	27	25	27	26	28	28	27	27	28	31	29	32	14	0	45	47	47	47
Wheat	43	43	46	45	45	44	44	44	47	46	46	46	46	46	48	49	45	45	0	54	47	50
Neurospora crassa	48	47	46	46	46	46	46	46	49	47	48	46	49	47	49	48	41	47	54	0	41	42
Baker's yeast	45	45	46	45	45	45	45	45	46	46	45	46	49	47	47	47	45	47	47	41	0	27
Candida krusei	51	51	51	50	50	49	50	50	51	51	50	51	53	51	51	48	47	47	50	42	27	0

[a]From M. O. Dayhoff (Ed.), *Atlas of Protein Sequence and Structure,* Volume 4, 1969. National Biomedical Research Foundation, Washington, D. C.

at the bottom of the tree. As mutational events occurred, the line of descent from the early ancestor must have diverged, eventually leading to the development of species typical of animals and plants. Those species with closely related cytochrome c sequences are shown close to one another on a main branch. For example, man and the monkey, the chicken and the penguin, the fruit fly and the moth, and the yeast and candida are all examples of closely related pairs. The evolutionary development of species as deduced from sequence studies of cytochrome c (and of other proteins) agrees with those proposed by the biologist based on classical taxonomic studies. Clearly, the chemical structure of proteins provides a reliable basis for extending these studies to include more species, to refine the analysis further, and to gain an insight into the possible sequences of a cytochrome c of a primitive organism.

Divergent evolution

We now turn to a group of enzymes with similarities in their biological function and mechanism of action, and we will examine the relatedness of their amino acid sequences. The enzymes that we will select for further study are the serine proteinases. The homologies that exist in the sequences of trypsin, chymotrypsin A, elastase, and thrombin (Fig. 11-9) strongly suggest that these proteins must be derived from a common precursor. The structure of the presumed precursor is not known, but one can nevertheless accept the idea that at one time an ancestoral species existed that contained a gene specifying the synthesis of a primitive type of proteolytic enzyme. The ancestoral gene was subject to random mutational events, and natural-selection pressures helped to propagate those changes that led to the development of an improved enzyme.

At some period in the reproduction of an organism, a special event occurred, namely, a gene duplication (Fig. 11-13). Gene duplication implies that two genes are now present in the organism, both capable of specifying the synthesis of the same proteolytic enzyme. Again, as a function of time, random mutational events operating on each gene would lead to their modification such that two distinct proteolytic enzymes could be synthesized by the organism. A repetition of gene doubling and subsequent alterations of each gene would produce new members of the group, each member containing numerous vestiges of their past history.

The development of new proteins from a common precursor is known as divergent evolution. The vestiges of the past found in present-day species are

---

Fig. 11-12. A phylogenetic tree depicting the evolution of cytochrome c. The numbers of inferred amino acid changes per 100 residues are shown on the branches of the tree. The branch lengths are related to the number of substitutions per 100 residues. The point of earliest time is placed on the lower part of the tree, shown by a dotted line. [From M. O. Dayhoff (Ed.), *Atlas of Protein Sequence and Structure,* Vol. 5. National Biomedical Research Foundation, Washington, D. C., 1972.]

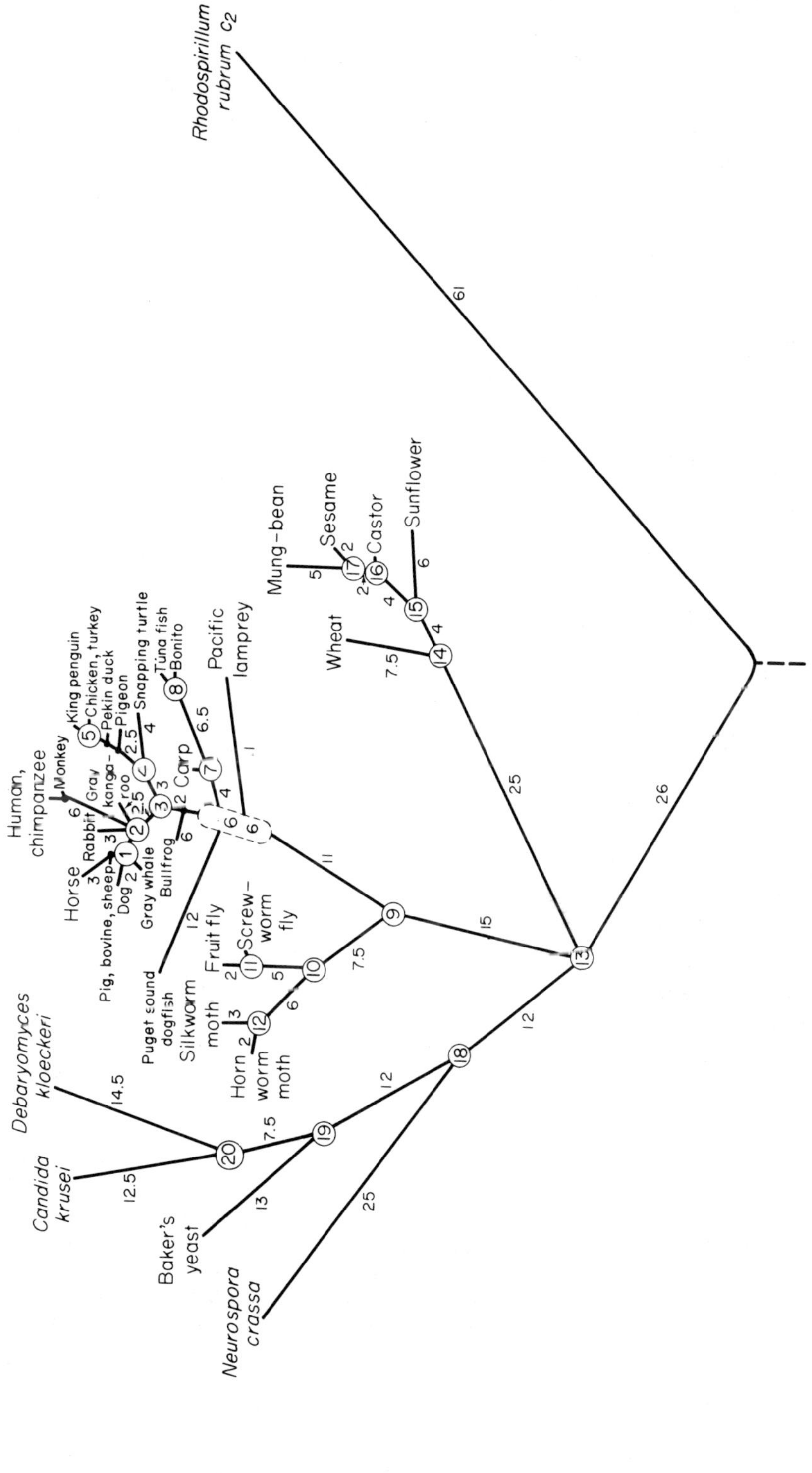

Rhodospirillum rubrum c_2
61
26
Mung-bean
Sesame
Castor
Sunflower
Wheat
25
Human, chimpanzee
Monkey
Horse
Rabbit
Gray kangaroo
Pig, bovine, sheep
Dog
Gray whale
King penguin
Chicken, turkey
Pekin duck
Pigeon
Snapping turtle
Tuna fish
Bonito
Carp
Bullfrog
Pacific lamprey
Puget sound dogfish
Fruit fly
Screw-worm fly
Silkworm moth
Horn worm moth
Debaryomyces kloeckeri
Candida krusei
Baker's yeast
Neurospora crassa
14.5
12.5
7.5
13
12
25
15
11

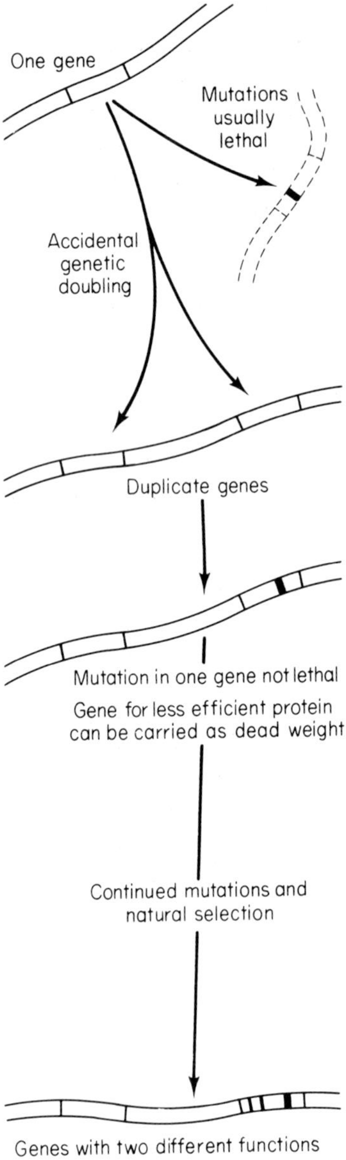

Fig. 11-13. The modification of a gene by a mutational event, the process of gene duplication, and the evolutionary process. [From R. E. Dickerson and I. Geis, *The Structure and Action of Proteins.* Harper and Row, New York, 1969.]

evident from the degree of homology of diverging proteins; if the homology is high, the structures of the pair are close to the structures that existed at the time of gene doubling. The order of appearance of each of the serine proteinases may become evident as the sequences of these enzymes become available from a number of different species.

Convergent evolution

In addition to divergent evolution specifying the similarity of proteins, two or more proteins may also be related by a process of convergent evolution. Consider the case of an ancestral organism with two unrelated genes, each capable of dictating the structure of a unique proteolytic enzyme. With time, each gene may be modified by mutational events in such a way that the two enzymes may now have similarities in their biological properties.

An example of convergent evolution is seen with the pair of proteolytic enzymes, chymotrypsin A and subtilisin; the former is the well-characterized enzyme of bovine origin, and the latter is elaborated by the bacterial organism, *Bacillis subtilis*. Both enzymes are classified as serine proteinases from chemical-modification studies, and both enzymes catalyze the hydrolysis of peptide bonds by a similar mechanism of action. In light of these functional similarities, it is remarkable that their amino acid sequences are completely different except for homologies of sequence at the active site. Furthermore, as expected, their three-dimensional structures differ except that both have the charge-relay system, the hydrogen-bonded alignment of the serine-histidine-aspartic acid (Fig. 11-6), as the arrangement of groups at the active site.

Convergent evolution is the pathway followed for the development of two dissimilar proteins that have converged on a common structural element for their active site. The evolutionary process takes advantage of those changes in structure that produce a repeating theme, namely, the construction of a group of side chain constituents properly oriented in space, which have catalytic activity. It would appear that both convergent and divergent evolution of enzymes describe methods of altering the sequence of a protein generally, while maintaining specific regions, those concerned with the catalytic events.

SUMMARY

In this concluding chapter we examined a number of structural features, typical of proteins generally and of proteolytic enzymes specifically, that relate to the biological activity of the molecule. We looked at methods that are suitable for the identification of those amino acid residues that constitute the active site of proteolytic enzymes; we also examined the role of these residues in the mechanism of catalytic action of the enzyme.

Sequence studies revealed a similarity, or homology, of structure of those proteins that are closely related with respect to their biological function. Homologous proteins have invariant residues—that is, identical residues—throughout their sequence, a situation suggesting that these residues act as critical structural elements and as components of the active center.

With the results obtained from x-ray studies of proteins at high resolution, it is now possible to correlate the sequence analysis of homologous proteins with the location of amino acid residues in the structure of the molecule. Most invariant residues are buried within the structure of globular proteins, and most variable residues are found on the surface. The buried residues participate in hydrophobic interactions, an important contribution to the stability of the molecule and an essential consideration in the proper folding of the polypeptide chain. The residues located on the surface are principally polar types, and they give to the structure their ability to interact with solvent molecules, with small organic structures, and with other proteins.

The correlation of structure with function allows us to appreciate a number of features typical of protein molecules. Although our understanding of protein structure is rather satisfying, it is clear that many new "rules of structure" are yet to be discovered. If our progress continues as well as it did in the recent past, the day will shortly arrive when the protein chemist can examine the sequence of a protein and predict its function as well as the molecule's three-dimensional structure.

REFERENCES

Books

M. L. Bender and L. J. Brubacher, *Catalysis and Enzyme Action.* McGraw-Hill, New York, 1973.

G. E. Means and R. E. Feeney, *Chemical Modification of Proteins.* Holden-Day, San Francisco, 1971.

S. A. Bernhard, *The Structure and Function of Enzymes.* Benjamin, New York, 1968.

Reviews and Articles

R. E. Dickerson, "The Structure and History of an Ancient Protein," *Sci. Am.*, p. 58 (April 1972).

R. E. Dickerson, "The Structure of Cytochrome c and the Rates of Molecular Evolution," *J. Molec. Evolution,* **1**, 26 (1971).

P. Jollès and J. Jollès, "Primary Sequences of Proteins and their Evolution," *Progress in Biophysics and Molecular Biology,* **22**, 127 (1971).

L. A. Cohen, "Chemical Modification as a Probe of Structure and Function," *The Enzymes,* **1**, 148 (1970).

H. Neurath, R. A. Bradshaw, and R. Arnon, "Homology and Phylogeny of Proteolytic Enzymes," in *Structure-Function Relationships of Proteolytic Enzymes,* P. Desnuelle, H. Neurath, and M. Otteson (Eds.). Academic Press, New York, 1970.

D. C. Phillips, D. M. Blow, B. S. Hartley, and G. Lowe (Organizers), "A Discussion on the Structures and Functions of Proteolytic Enzymes," *Phil. Trans. Royal Soc. London, B.*, **257**, 63-266 (1970).

E. Scoffone and A. Fontana, "Identification of Specific Amino Acid Residues," in *Protein Sequence Determination*, S. B. Needleman (Ed.). Springer-Verlag, New York, 1970.

E. Shaw, "Chemical Modification by Active-Site-Directed Reagents," *The Enzymes*, **1**, 91 (1970).

E. L. Smith, "Evolution of Enzymes," *The Enzymes*, **1**, 267 (1970).

G. R. Stark, "Recent Developments in Chemical Modification and Sequential Degradation of Proteins," *Adv. Prot. Chem.*, **24**, 261 (1970).

M. O. Dayhoff, "Computer Analysis of Protein Evolution," *Sci. Am.*, **221**, 86 (July 1969).

L. A. Cohen, "Group-Specific Reagents in Protein Chemistry," *Annual Rev. Biochem.*, **37**, 695 (1968).

S. J. Singer, "Covalent Labeling of Active Sites," *Adv. Prot. Chem.*, **22**, 1 (1967).

E. Margoliash and A. Schejter, "Cytochrome c," *Adv. Prot. Chem.*, **21**, 144 (1966).

D. C. Phillips, "The Three-Dimensional Structure of an Enzyme Molecule," *Sci. Am.*, **215**, 78 (November 1966).

G. H. Dixon, "Mechanisms of Protein Evolution," in *Essays in Biochemistry*, P. N. Campbell and G. D. Greville (Eds.). Academic Press, New York, **2**, 147 (1966).

E. Zuckerkandl and L. Pauling, "Potential Value of Proteins for Providing a Document of a Species' Evolutionary History," *J. Theoret. Biol.*, **8**, 357 (1965).

INDEX